Trigonometry

Xing Zhou

Math for Gifted Students

http://www.mathallstar.org

Copyright © 2015 by Xing Zhou. All rights reserved.

No part of this book may be reproduced, distributed or transmitted in any form or by any means, including photocopying, scanning, or other electronic or mechanical methods, without written permission of the author.

To promote education and knowledge sharing, reuse of some contents of this book may be permitted, courtesy of the author, provided that: (1) the use is reasonable; (2) the source is properly quoted; (3) the user bears all responsibility, damage and consequence of such use. The author hereby explicitly disclaims any responsibility and liability; (4) the author is notified in advance; and (5) the author encourages, but does not enforce, the user to adopt similar policies towards any derived work based on such use.

Please visit the website `http://www.mathallstar.org` for more information or email `contact@mathallstar.org` for suggestions, comments, questions and all copyright related issues.

use your mobile device to scan this QR code for more resources including books, practice problems, online courses, and blog.

This book was produced using the LaTeX system.

Contents

1 Introduction   1

2 **Trigonometry Basics**   3
    2.1 Trigonometric Functions Defined 3
    2.2 Special Angles 5
    2.3 Tackling $(k \cdot 90° \pm \alpha)$ 8
    2.4 Trigonometric Function Properties 10
    2.5 Trigonometric Inequality 12
    2.6 Additional Trigonometric Functions 14
    2.7 Practice 15

3 **Trigonometric Identities**   19
    3.1 Sum and Difference of Angles 19
    3.2 Double and Half Angle 22
    3.3 Triple Angle 25
    3.4 Product-Sum and Sum-Product 27
    3.5 Practice 31

4 **Trigonometric Techniques**   35
    4.1 $a \sin \theta + b \cos \theta = \sqrt{a^2 + b^2} \sin(\theta + \varphi)$ 35
    4.2 Multiplying and Telescoping 37
    4.3 Reverse Construction 42
    4.4 Symmetry and Pairing 44
    4.5 Mathematical Induction 45
    4.6 Half-Tangent Substitution 46
    4.7 Practice 47

5 **The Complex Number Method**   51
    5.1 Trigonometry to Complex Number 51
    5.2 Examples 52
    5.3 Practice 57

6 **Trigonometry in Triangle**   61
    6.1 Law of Sines, Cosines and Tangents 61

CONTENTS

    6.2   Areas, Circumradius and Inradius 64
    6.3   Trigonometric Ceva's Theorem 65
    6.4   Triangular Trigonometric Identities 66
    6.5   Triangular Trigonometric Inequalities 69
    6.6   Practice 71

7 Additional Techniques   **75**
    7.1   Applying Law of Cosines 75
    7.2   Substituting $(a^2 \pm b^2 = r^2)$ 78
    7.3   Substitution by $\tan \theta$ 78
    7.4   Converse of Triangular Identities 80
    7.5   Practice 82

8 Solutions   **85**
    8.1   Introduction 85
    8.2   Trigonometry Basics 86
    8.3   Trigonometric Identities 96
    8.4   Trigonometric Techniques 108
    8.5   The Complex Number Method 118
    8.6   Trigonometry in Triangle 124
    8.7   Additional Techniques 134

Preface

Welcome to Math All Star© series!

Math All Star originates from a series of lectures given to a group of gifted middle school students with a love for mathematics and an interest in participating in competitions such as MathCounts, AMC, and AIME. These lectures aim to strengthen their problem-solving abilities and to introduce effective techniques that are not typically taught in the classroom.

As the popularity of Math All Star grew, the author began to upload lecture materials to create online courses, thereby providing students with the opportunity to progress at their own paces.

Since then, course materials have constantly been reviewed and updated to reflect student feedback and the observations made during lectures. Recent competition problems are also continuously analyzed and referenced to ensure the relevance of the contents. These course materials are the foundations of this Math All Star series.

Because competition math is a diversified subject that covers both a wide breadth and depth of topics, it is quite challenging to effectively cover all the material in one book that is appropriate for every interested student. Consequently, the author has decided to write a series of books, with each one focusing on a particular topic. Students are encouraged to pick and choose where to begin, depending on their individual skill levels and needs.

In addition to these books, the Math All Star website provides extra practice problems and serves as a highly recommended supplemental learning resource.

If there are any questions, comments, or concerns, please visit the website or email `contact@mathallstar.org`.

Happy learning!

To visit the Math All Star website, scan this QR code or go directly to
http://www.mathallstar.org

Chapter 1

Introduction

Trigonometry is an important subject in mathematics. It relates to many other subjects such as geometry, coordinate geometry, complex number, and so on. We see trigonometric problems appear in almost every AMC12 or above competition either explicitly or implicitly. In addition, students attending lower level competition may find trigonometry can offer valuable alternative solutions to some geometry problems.

In order to be proficient in trigonometry, it is necessary to memorize formulas. However, there are hundreds, if not thousands, of trigonometric formulas. It is practically impossible and often unnecessary to remember all of them. Therefore, it is critical to know what formulas are essential and thus have to be remembered. Accordingly, the first objective of this book is to help students understand and remember those essential formulas.

Remembering a sufficient number of formulas may help students achieve high scores in school tests. However, it is not sufficient to win math competitions. Students will have to master relevant techniques and be able to choose the most appropriate formula to solve a particular problem. Let's take the following expression as

Chapter 1: Introduction

an example:
$$\cos 20° \cos 40° \cos 80° \qquad (1.1)$$

The value of this expression can be calculated in multiple ways. A classic technique is to multiply it by $\sin 20°$. The result can be obtained by applying the double angle formula a few times. An alternative, relatively less known, solution is to apply the triple angle formula. This solution can produce the result immediately. Both approaches are workable in this case. Each of them can be used to tackle some generalized formed of *(1.1)*. As such, it is important for students to know all the relevant techniques and which one to choose in a particular case. Hence, the second objective of this book is to illustrate important techniques and to explain when to use them. In order to achieve this, some sample problems will appear repeatedly when different techniques are discussed. This will help students understand the pros and cons of different techniques in tackling specific problems.

Upon completing this book, students should have the necessary basis for solving trigonometry problems in math competitions. In order to maximize learning results, students should attempt all the examples and practice problems once again after finishing the whole book. This will be helpful to re-enforce those techniques discussed and also offer a chance for students to reflect appropriateness of different techniques in solving particular problems.

Chapter 2

Trigonometry Basics

2.1 Trigonometric Functions Defined

Trigonometric functions can be defined using the ratios of side lengths of a right triangle.

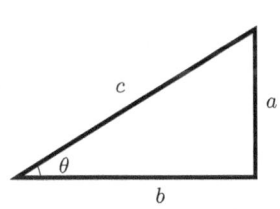

$$\sin\theta = \frac{opposite}{hypotenuse} = \frac{a}{c}$$

$$\cos\theta = \frac{adjacent}{hypotenuse} = \frac{b}{c}$$

$$\tan\theta = \frac{opposite}{adjacent} = \frac{a}{b}$$

These are the three most used trigonometric functions. Additional ones will be introduced later.

Based on these definitions, it is clear that:

$$\tan\theta = \frac{\sin\theta}{\cos\theta} \qquad (2.1)$$

and

$$\sin(90° - \theta) = \cos\theta \quad , \quad \cos(90° - \theta) = \sin\theta \qquad (2.2)$$

Chapter 2: Trigonometry Basics

Meanwhile, it can be shown that

$$\sin^2 \theta + \cos^2 \theta = 1 \qquad (2.3)$$

by replacing these two trigonometric functions with their definitions and then applying the Pythagorean theorem. In fact, (2.3) can be regarded as Pythagorean theorem's trigonometric form. It is one of the most important and basic trigonometric relationships.

While defining trigonometric functions using a right triangle is convenient, it limits the angle to be acute. This restriction can be eliminated by using coordinate grid based definition.

Let's take a point P, which can be any point but the origin, on the ray whose included angle with the X axis is θ. Then, all the trigonometric functions can be defined using P's coordinates x and y, as shown below where $r = \sqrt{x^2 + y^2}$ is the distance from P to the origin. The values of these ratios remain the same when P travels along the ray because of similar triangle's properties.

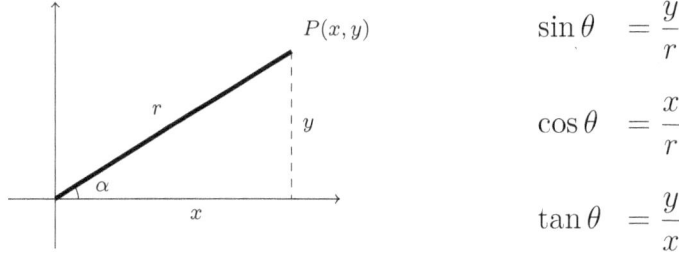

With this definition, θ can take any value including negative ones and those greater than 360°.

Domains for both sine and cosine functions are all real numbers, i.e. the entire $\mathbb{R}$. However, the tangent function is undefined when $x = 0$, or equivalently, $\theta = \frac{k\pi}{2}$ where k is an odd integer. Consequently, the domain for tangent function is all real numbers excluding these points.

Meanwhile, though r is always positive, both x and y can be positive, negative or zero. Therefore, the ranges of sine and cosine functions are

$$-1 \leq \sin\theta \leq 1 \quad , \quad -1 \leq \cos\theta \leq 1 \tag{2.4}$$

because $|x| \leq r$ and $|y| \leq r$.

It is obvious that the ratio of x and y can be any value. As a result, the tangent function's range includes all real numbers, i.e.,

$$\tan\theta \in \mathbb{R} \tag{2.5}$$

One implication of (2.5) is that, given any real number x, it is always possible to find an angle θ satisfying $x = \tan\theta$. Therefore, an algebraic expression with respective to x can be transformed to a trigonometric expression with respect to θ. This is an advanced and powerful technique which will be discussed later in this book.

2.2 Special Angles

For every angle θ, its trigonometric values are uniquely determined. However, it is unnecessary to know and remember all of them. All trigonometry problems which appear in both school tests and math competitions can be solved by applying appropriate transformations on a handful of special angles' trigonometric values. This section discusses these must-know special angles.

Firstly, it is only necessary to investigate angles between 0° and 90°. This is because trigonometric values of those angles falling outside this region can be obtained by corresponding ones within this region. Such transformation will be discussed in *Section 2.3*.

Those absolutely must-know special angles even for school tests are listed below. Their trigonometric values can all be calculated using basic geometry.

Chapter 2: Trigonometry Basics

$x =$	$0°$	$30°$	$45°$	$60°$	$90°$
$\sin(x)$	0	$\frac{1}{2}$	$\frac{\sqrt{2}}{2}$	$\frac{\sqrt{3}}{2}$	1
$\cos(x)$	1	$\frac{\sqrt{3}}{2}$	$\frac{\sqrt{2}}{2}$	$\frac{1}{2}$	0
$\tan(x)$	0	$\frac{\sqrt{3}}{3}$	1	$\sqrt{3}$	−

Additional special angles which are necessary for competing in math contests include 15° and 18°. Their trigonometric values are listed below

$$\sin 15° = \frac{\sqrt{6}-\sqrt{2}}{4} \quad , \quad \cos 15° = \frac{\sqrt{6}+\sqrt{2}}{4} \qquad (2.6)$$

and

$$\sin 18° = \frac{\sqrt{5}-1}{4} \qquad (2.7)$$

The value of $\cos 18° = \sqrt{\frac{5}{8}+\frac{\sqrt{5}}{5}}$ which can be computed using (2.3) and (2.7) is relatively less used.

Values given in (2.6) and (2.7) can also be computed using geometry. For example, because 15° is half of 30° and 30°−60°−90° is a special right triangle in geometry, both $\sin 15°$ and $\cos 15°$ can be calculated using the angle bisector and Pythagorean theorem. Calculating these two values will be left as a practice problem. Computing $\sin 18°$ is relatively more challenging. Its solution is given in the next example.

Example 2.2.1

Compute the value of $\sin 18°$.

Solution

Let's construct $\triangle ABC$ where $\angle A = 36°$ and $\angle B = \angle C = 72°$. Assuming the bisector of $\angle B$ meets AC at point D. Then, $\angle ABD =$

Chapter 2: Trigonometry Basics

$\angle DBC = 36°$. It is easy to see that $\triangle ABC$, $\triangle DAB$ and $\triangle BCD$ are all isosceles.

Meanwhile, let H be the foot of the altitude drawn from vertex A. Because $\triangle ABC$ is isosceles and $\angle A = 36°$, we find $\angle BAH = \angle CAH = 18°$. Then, our objective is to compute

$$\sin 18° = \frac{BH}{AB}$$

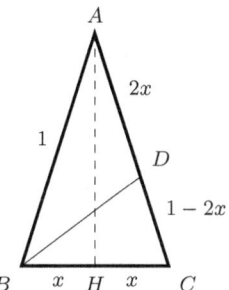

For convenience, let's assume $AB = 1$ and $BH = HC = x$. Then, because $\triangle BCD$ and $\triangle DAB$ are both isosceles, it must be true that
$$AD = DB = BC = 2x$$
Accordingly,
$$CD = AC - AD = 1 - 2x$$

Meanwhile, we claim that $\triangle ABC \sim \triangle BCD$ because both of them are $36° - 72° - 72°$ triangles.

$$\therefore \quad \frac{AB}{BC} = \frac{BC}{CD} \implies \frac{1}{2x} = \frac{2x}{1-2x} \implies x = \frac{\sqrt{5}-1}{4}$$

Now, it follows that

$$\sin 18° = \frac{BH}{AB} = \frac{x}{1} = \frac{\sqrt{5}-1}{4}$$

Done.

The 18° angle is closely related to a regular pentagon. This is because all interior angles of a regular pentagon are 108° which equals

Chapter 2: Trigonometry Basics

(90° + 18°). Half of an interior angle equals 54° which is three times of 18°. Trigonometric values of both 108° and 54° can be computed exactly using those of 18°. Therefore, regular pentagon related geometry problems may be solved using straightforward trigonometric computation.

2.3 Tackling $(k \cdot 90° \pm \alpha)$

If the trigonometric value of $\alpha \in [0, \frac{\pi}{2}]$ is known, then it is possible to obtain trigonometric values of angle $\beta = (k \cdot 90° \pm \alpha)$ where k is an integer.

Let's illustrate this using $\sin \beta$ as an example. Values of other functions, such as $\cos \beta$, can be derived in a similar way.

In order to obtain $\sin \beta$, we must determine two things:

i) The sign of $\sin \beta$: it can be either positive or negative,

ii) The absolute value of $\sin \beta$: it can be either $\sin \alpha$ or $\cos \alpha$.

The sign of $\sin \beta$ is decided by the quadrant where β locates. By definition, $\sin \beta = \frac{y}{r}$. As r is always positive, the sign of $\sin \beta$ is the same as the sign of y. Hence, it is positive in the I and II quadrants and negative in the III and IV quadrants.

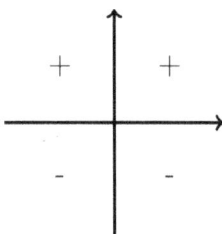

The absolute value of $\sin \beta$ can be determined by observing the result of rotating α. When angle α is rotated by a multiple k of

Chapter 2: Trigonometry Basics

90°, the original triangle (as seen in the following diagrams) will either be still horizontal if k is even, or become vertical if k is odd. When the triangle settles as horizontal, the lengths of the opposite side and the adjacent side are still the same as those of the original triangle. However, when the triangle becomes vertical, they will switch.

Consequently, when k is even:

$$|\sin\beta| = \frac{|y|}{r} = |\sin\alpha|$$

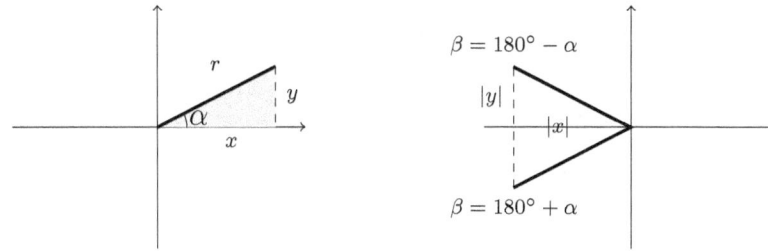

When k is odd, the opposite side is the original adjacent one:

$$|\sin\beta| = \frac{|x|}{r} = |\cos\alpha|$$

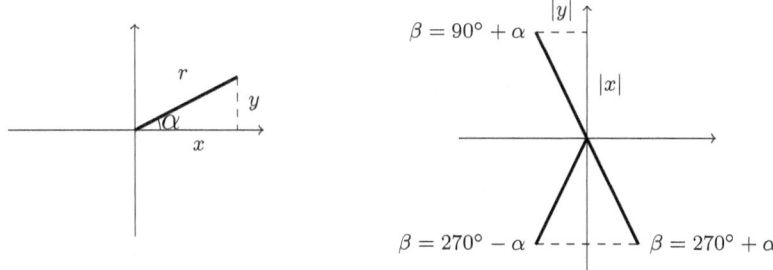

Upon having decided both the sign and the absolute value of $\sin\beta$, we can determine the exact value of $\sin\beta$.

Let's consider an example.

Chapter 2: Trigonometry Basics

Example 2.3.1

Compute $\sin 150°$, $\sin 300°$, and $\cos 240°$.

Solution

Because $\sin 30° = \frac{1}{2}$ and $\cos 30° = \frac{\sqrt{3}}{2}$, therefore,

$$\sin 150° = \sin(180° - 30°) = +|\sin 30°| = \frac{1}{2}$$

$$\sin 300° = \sin(270° + 30°) = -|\cos 30°| = -\frac{\sqrt{3}}{2}$$

$$\cos 240° = \cos(270° - 30°) = -|\sin 30°| = -\frac{1}{2}$$

Done.

2.4 Trigonometric Function Properties

Because the ending rays of angles θ and $(\theta + 2\pi)$ are the same, the values of all the trigonometric functions of θ and $(\theta + 2\pi)$ must be equal. This means that all trigonometric functions are periodical functions. Although 2π must be a qualified period for all trigonometric functions, it may not be the smallest one. For example, while the periods of both sine and cosine functions are 2π, the tangent function's period is π.

$$\sin(\theta + 2\pi) = \sin\theta \quad , \quad \cos(\theta + 2\pi) = \cos\theta \qquad (2.8)$$

but

$$\tan(\theta + \pi) = \tan\theta \qquad (2.9)$$

These can be clearly seen in their respective function plots.

Chapter 2: Trigonometry Basics

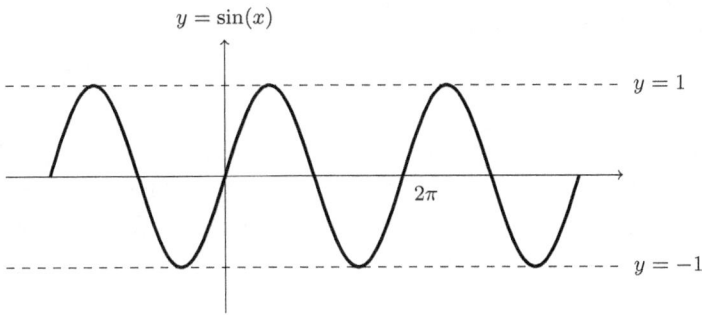

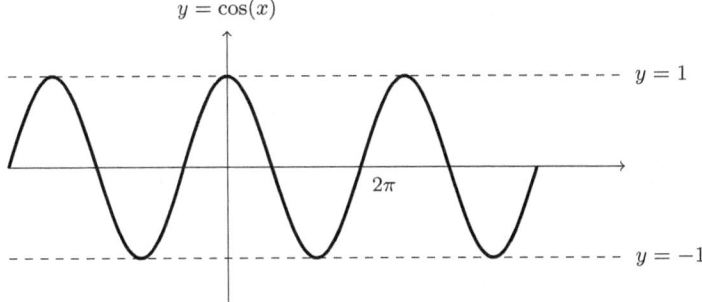

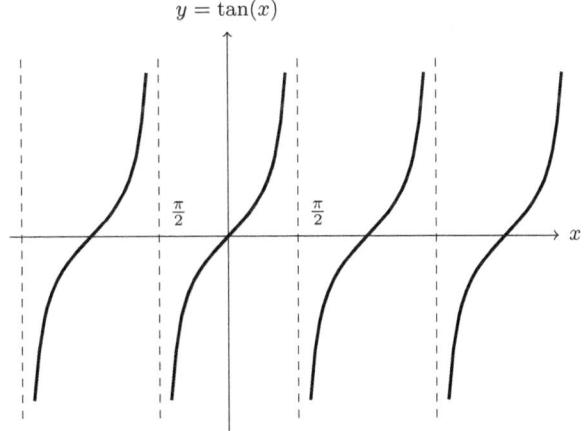

From these plots, it is also clear that sine and tangent are odd functions because their graphs are symmetric with respect to the origin, and cosine function is even because its graph is symmetric

Chapter 2: Trigonometry Basics

with respect to the y axis.

$$\sin(-x) = -\sin x, \cos(-x) = \cos x, \tan(-x) = -\tan x \quad (2.10)$$

In addition, all these functions have periodically repeated increasing and decreasing regions. These pattern can be observed from their function plots. For example, the sine function is monotonically increasing in $[-90°, 90°]$ and is monotonically decreasing in $[90°, 270°]$.

Despite of appearing to be simple and intuitive, these properties can play critical roles in solving some interesting competition problems. An example will be given in the next section.

2.5 Trigonometric Inequality

Relations (2.4), which is copied below, is the most basic trigonometric inequality.

$$-1 \leq \sin x, \cos x \leq 1$$

Additionally, competition participants should also be familiar with the following inequality where x is expressed in radian.

$$\sin x < x < \tan x, \quad \left(0 < x < \frac{\pi}{2}\right) \quad (2.11)$$

This inequality can be explained using the diagram below.

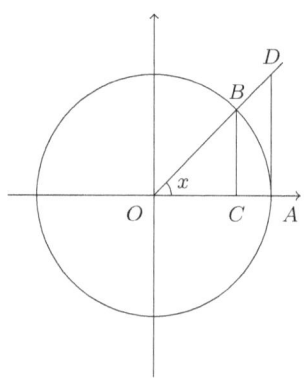

Chapter 2: Trigonometry Basics

Let O be a unit circle which means its radius $OA = OB = 1$. If $\angle DOA = x$ (in radian), then arc $\widehat{AB} = x$ by the definition of radian measurement. Meanwhile, based on the definitions of trigonometric functions, we find segment $BC = \sin x$, and $DA = \tan x$.

It is intuitive to see that $BC < \widehat{AB} < AD$ which is equivalent to asserting $\sin x < x < \tan x$.

When a trigonometric inequality involves both a real number x and its trigonometric function values, it usually can be solved by combining *(2.11)* and trigonometric functions' properties which are discussed in the previous section. Let's review an example.

Example 2.5.1

If $0 < x < \frac{\pi}{4}$, arrange $\sin x$, $\sin(\sin x)$, and $\sin(\tan x)$ in an ascending order.

Solution

Let's first compare $\sin(\sin x)$ and $\sin x$. Their relation can be determined using *(2.11)* by replacing x with $\sin x$. The only thing we need to make sure is that $\sin x$ falls in the region of $\left(0, \frac{\pi}{2}\right)$. This is indeed true because sine function is monotonically increasing between 0 and $\frac{\pi}{4}$. Consequently,

$$\sin 0 < \sin x < \sin \frac{\pi}{4} \implies 0 < \sin x < \frac{\sqrt{2}}{2} < \frac{\pi}{2}$$

Hence, we find
$$\sin(\sin x) < \sin x \qquad (2.12)$$

Now, let's determine the relative values of $\sin x$ and $\sin(\tan x)$. Applying *(2.11)* again yields

$$0 < x < \frac{\pi}{4} \implies x < \tan x$$

Because tangent function is also increasing in $(0, \frac{\pi}{4})$, we find

$$0 = \tan 0 < \tan x < \tan \frac{\pi}{4} = 1 < \frac{\pi}{2}$$

This means
$$0 < x < \tan x < \frac{\pi}{2}$$
Now, the fact that sine function is increasing in this region means
$$\sin x < \sin(\tan x) \qquad (2.13)$$

Combining both *(2.12)* and *(2.13)* leads to the conclusion
$$\boxed{\sin(\sin x) < \sin x < \sin(\tan x)}$$

Done.

2.6 Additional Trigonometric Functions

In addition to sine, cosine and tangent, there are three other trigonometric functions which also appear frequently.

$$\cot \alpha = \frac{1}{\tan \alpha} = \frac{\cos \alpha}{\sin \alpha} \qquad (2.14)$$
$$\sec \alpha = \frac{1}{\cos \alpha} \qquad (2.15)$$
$$\csc \alpha = \frac{1}{\sin \alpha} \qquad (2.16)$$

Based on these definitions, it can be shown that
$$\sec^2 \alpha = 1 + \tan^2 \alpha \qquad (2.17)$$
$$\csc^2 \alpha = 1 + \cot^2 \alpha \qquad (2.18)$$

Problems related to these three functions can be either solved directly, or be transformed to sine, cosine, and tangent functions first.

2.7 Practice

Practice 1

Compute the following values:

    i) $\cos 75°$

    ii) $\sin 165°$

    iii) $\sin 105°$

Practice 2

Show that
$$\sec^2 \alpha = 1 + \tan^2 \alpha$$
$$\csc^2 \alpha = 1 + \cot^2 \alpha$$

Practice 3

Prove the identity: $\tan^2 x - \sin^2 x = \tan^2 x \sin^2 x$.

Practice 4

Compute $\sin 15°$ and $\cos 15°$ using a geometry approach.

Chapter 2: Trigonometry Basics

Practice 5

Let a and b be the two sides of a triangle, and their included angle be C. Show that the area of this triangle equals

$$S = \frac{1}{2} \cdot ab \sin C \tag{2.19}$$

Practice 6

If $\cos x - \sin x = \sqrt{2} \sin x$, prove $\cos x + \sin x = \sqrt{2} \cos x$.

Practice 7

Find an acute angle α so that

$$\sqrt{15 - 12\cos\alpha} + \sqrt{7 - 4\sqrt{3}\sin\alpha} = 4$$

Practice 8

Find the range of real number a if the following equation of x is solvable in real numbers:

$$\sin^2 x + \cos x + a = 0$$

Practice 9

Find the number of solutions to the equation $\sin x = \frac{x}{2018}$.

Practice 10

Prove that function $f(x) = \cos \sqrt{x}$ is not a periodical function.

Practice 11

Sort $\sin(-1)$, $\cos(-1)$, and $\tan(-1)$ in an ascending order.

Practice 12

Let $0° < \alpha < 45°$, explain that

$$(\tan \alpha)^{\cot \alpha} < (\tan \alpha)^{\tan \alpha} < (\cot \alpha)^{\tan \alpha} < (\cot \alpha)^{\cot \alpha}$$

Practice 13

Let $\theta \in [0, 2\pi]$ satisfying

$$\cos^5 \theta - \sin^5 \theta < 7(\sin^3 \theta - \cos^3 \theta)$$

Find the range of θ.

Practice 14

If $0 < \alpha < \beta < \frac{\pi}{2}$, show

$$\frac{\cot \beta}{\cot \alpha} < \frac{\cos \beta}{\cos \alpha} < \frac{\beta}{\alpha}$$

Chapter 2: Trigonometry Basics

Chapter 3

Trigonometric Identities

In order to be proficient in solving trigonometry problems, one must be familiar with some identities in addition to those basic definitions introduced in the preceding chapter. These identities are fundamental formulas which deal with basic trigonometric operations such as adding or subtracting two angles. In fact, a great amount of trigonometric problems can be solved by applying these basic identities.

3.1 Sum and Difference of Angles

Trigonometric functions are not linear. This means that $\sin(\alpha \pm \beta) \neq \sin \alpha \pm \sin \beta$. Instead, the following formulas must be used when adding or subtracting two angles. They are the basis to derive other trigonometric identities.

$$\sin(\alpha \pm \beta) = \sin \alpha \cos \beta \pm \cos \alpha \sin \beta \qquad (3.1)$$

$$\cos(\alpha \pm \beta) = \cos \alpha \cos \beta \mp \sin \alpha \sin \beta \qquad (3.2)$$

$$\tan(\alpha \pm \beta) = \frac{\tan \alpha \pm \tan \beta}{1 \mp \tan \alpha \tan \beta} \qquad (3.3)$$

Chapter 3: Trigonometric Identities

Proof of these and other basic identities can be found in many textbooks or online. Thus, for concise reason, not all proofs will be given in this book. Only those involving typical techniques will be discussed.

Example 3.1.1

Prove the sum of angle formula.

$$\sin(\alpha + \beta) = \sin\alpha \cos\beta + \cos\alpha \sin\beta \tag{3.4}$$

Proof

One way to prove this is to use the area method[1].

It is sufficient to only discuss the case where $0° \leq \alpha, \beta < 90°$. This is because when angles fall outside this region, it is always possible to transform them into equivalent ones within this region by adding or subtracting a multiple of 90°. Tackling angle in the form of $(k \cdot 90° \pm \alpha)$ is discussed in *Section 2.3* on *page 8*.

When at least one of the two angles equals 0°, *(3.4)* obviously holds. Therefore, we only need to consider the case when both α and β are acute. In such a case, it is always possible to construct a $\triangle ABC$ as shown below where AD is the altitude.

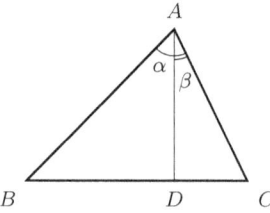

---

[1]The area method is discussed in the book *Geometry Techniques* written by the same author.

Chapter 3: Trigonometric Identities

Applying the triangle's area formula[2]

$$S = \frac{1}{2} \cdot ab \sin C$$

on $S_{\triangle ABC}$, $S_{\triangle ABD}$, and $S_{\triangle ACD}$, respectively, yield:

$$S_{\triangle ABC} = S_{\triangle ABD} + S_{\triangle ACD}$$
$$\frac{1}{2} \cdot AB \cdot AC \cdot \sin(\alpha + \beta) = \frac{1}{2} \cdot AB \cdot AD \cdot \sin\alpha + \frac{1}{2} \cdot AC \cdot AD \cdot \sin\beta$$
$$\sin(\alpha + \beta) = \sin\alpha \cdot \frac{AD}{AC} + \sin\beta \cdot \frac{AD}{AB}$$
$$\sin(\alpha + \beta) = \sin\alpha \cos\beta + \sin\beta \cos\alpha$$

QED

One basic application of the sum and difference angle formulas is to compute trigonometric values of those angles which can be written as the sum or the difference of two special angles whose trigonometric values are known. Let's reconsider the calculation of $\sin 15°$ and $\cos 15°$ here.

Example 3.1.2

Compute the value of $\sin 15°$ and $\cos 15°$.

Solution

Writing $15° = 45° - 30°$ gives

$$\sin 15° = \sin(45° - 30°)$$
$$= \sin 45° \cos 30° - \cos 45° \sin 30°$$
$$= \frac{\sqrt{2}}{2} \cdot \frac{\sqrt{3}}{2} - \frac{\sqrt{2}}{2} \cdot \frac{1}{2}$$
$$= \boxed{\frac{\sqrt{6} - \sqrt{2}}{4}}$$

---

[2]This formula is one of the practice problems in the previous chapter.

Chapter 3: Trigonometric Identities

$$\begin{aligned}
\cos 15° &= \cos(45° - 30°) \\
&= \cos 45° \cos 30° + \sin 45° \sin 30° \\
&= \frac{\sqrt{2}}{2} \cdot \frac{\sqrt{3}}{2} + \frac{\sqrt{2}}{2} \cdot \frac{1}{2} \\
&= \boxed{\frac{\sqrt{6} + \sqrt{2}}{4}}
\end{aligned}$$

Done.

It is easy to verify that $(\sin^2 15° + \cos^2 15°)$ indeed equals 1. In fact, when one of the sine and the cosine values is known, the other can be computed using *(2.3)* on *page 4*.

A useful variant of the sum of tangents formula is given below:

$$\tan(\alpha \pm \beta) = \frac{\tan \alpha \pm \tan \beta}{1 \mp \tan \alpha \tan \beta}$$
$$\implies \tan \alpha \pm \tan \beta = \tan(\alpha \pm \beta)(1 \mp \tan \alpha \tan \beta) \quad (3.5)$$

This identity can often be used when the sum of two tangents is involved, especially when the sum is a special angle.

3.2 Double and Half Angle

A special case of the sum formula is that the two angles are equal. Setting $\beta = \alpha$ in *(3.1)* to *(3.3)* on *page 19* gives the double angle formulas:

$$\sin(2\alpha) = 2\sin\alpha\cos\alpha \quad (3.6)$$
$$\cos(2\alpha) = \cos^2\alpha - \sin^2\alpha = 2\cos^2\alpha - 1 = 1 - 2\sin^2\alpha \quad (3.7)$$
$$\tan(2\alpha) = \frac{2\tan\alpha}{1 - \tan^2\alpha} \quad (3.8)$$

Chapter 3: Trigonometric Identities

Among these three formulas, *(3.7)* is an interesting one because it can relate $\cos(2\alpha)$ with either $\cos\alpha$ alone or $\sin\alpha$ alone. Hence, replacing α with $\frac{1}{2}\cdot\alpha$ leads to the half angle formula:

$$\sin^2\frac{\alpha}{2} = \frac{1-\cos\alpha}{2} \implies \sin\frac{\alpha}{2} = \pm\sqrt{\frac{1-\cos\alpha}{2}} \qquad (3.9)$$

$$\cos^2\frac{\alpha}{2} = \frac{1+\cos\alpha}{2} \implies \cos\frac{\alpha}{2} = \pm\sqrt{\frac{1+\cos\alpha}{2}} \qquad (3.10)$$

Signs in these two formulas before the radical operators depend on which quadrant the half angle $\frac{\alpha}{2}$ locates. For a given angle, the result is unique, i.e. either positive or negative, but not both.

Tangent of half angle can be obtained by applying *(2.1)* on *(3.9)* and *(3.10)*. This is shown in the next example.

$$\tan\frac{\alpha}{2} = \pm\sqrt{\frac{1-\cos\alpha}{1+\cos\alpha}} = \pm\frac{\sin\alpha}{1+\cos\alpha} = \pm\frac{1-\cos\alpha}{\sin\alpha} \qquad (3.11)$$

While the first result in *(3.11)* is obvious, derivation of the 2nd and the 3rd require a few steps.

Example 3.2.1

Show that

$$\sqrt{\frac{1-\cos\alpha}{1+\cos\alpha}} = \pm\frac{\sin\alpha}{1+\cos\alpha} = \pm\frac{1-\cos\alpha}{\sin\alpha}$$

Proof

When $(1\pm\cos\alpha)$ or $(1\pm\sin\alpha)$ appears, the identity *(2.3)* on page 4 is always a candidate for consideration.

$$\sqrt{\frac{1-\cos\alpha}{1+\cos\alpha}} = \sqrt{\frac{(1-\cos\alpha)(1+\cos\alpha)}{(1+\cos\alpha)(1+\cos\alpha)}} = \sqrt{\frac{1-\cos^2\alpha}{(1+\cos\alpha)^2}}$$

Chapter 3: Trigonometric Identities

$$= \sqrt{\frac{\sin^2 \alpha}{(1+\cos\alpha)^2}} = \pm\frac{\sin\alpha}{1+\cos\alpha}$$

and

$$\sqrt{\frac{1-\cos\alpha}{1+\cos\alpha}} = \sqrt{\frac{(1-\cos\alpha)(1-\cos\alpha)}{(1+\cos\alpha)(1-\cos\alpha)}} = \sqrt{\frac{(1-\cos\alpha)^2}{1-\cos^2\alpha}}$$

$$= \sqrt{\frac{(1-\cos\alpha)^2}{\sin^2\alpha}} = \pm\frac{1-\cos\alpha}{\sin\alpha}$$

QED

Half angle formulas often associate with nested radical expressions. Simplifying nested radical expressions is discussed in the book *Power Calculation by Examples* written by the same author.

Let's consider an example.

Example 3.2.2

Calculate the exact value of $\sin 15°$ using the half angle formula.

Solution

Given $15°$ is half of $30°$, we have

$$\sin 15° = \sqrt{\frac{1-\cos 30°}{2}} = \sqrt{\frac{1-\frac{\sqrt{3}}{2}}{2}} = \sqrt{\frac{2-\sqrt{3}}{4}}$$

$$= \sqrt{\frac{8-4\sqrt{3}}{16}} = \frac{\sqrt{(\sqrt{6})^2 + (\sqrt{2})^2 - 2\sqrt{6}\cdot 2}}{4}$$

$$= \frac{\sqrt{6}-\sqrt{2}}{4}$$

Done.

It is clear that using the difference of angle formulas is easier than using the half angle formulas to compute sin 15° even though both formulas can do the job. This suggests that the ability to choose the most appropriate formulas to solve a particular problem is an important skill to master. More examples to illustrate this point will be presented later.

3.3 Triple Angle

Although triple angle formulas are usually not taught in classrooms, they are must-know for competition participants. This is because many competition trigonometric problems can be solved using these formulas directly.

$$\sin(3\alpha) = 3\sin\alpha - 4\sin^3\alpha \qquad (3.12)$$

$$\cos(3\alpha) = 4\cos^3\alpha - 3\cos\alpha \qquad (3.13)$$

$$\tan(3\alpha) = \frac{3\tan\alpha - \tan^3\alpha}{1 - 3\tan^2\alpha} \qquad (3.14)$$

One obvious way to prove these formulas is to first expand using $3\alpha = (2\alpha + \alpha)$ and then further expand using the double angle formula. Let's take the proof of *(3.12)* as example.

Example 3.3.1

Show that $\sin 3\alpha = 3\sin\alpha - 4\sin^3\alpha$.

Proof

By *(3.1)*, *(3.6)* etc, we have

$$\begin{aligned}
\sin 3\alpha &= \sin(2\alpha + \alpha) \\
&= \sin 2\alpha \cos\alpha + \cos 2\alpha \sin\alpha \\
&= (2\sin\alpha\cos\alpha)\cos\alpha + (1 - 2\sin^2\alpha)\sin\alpha
\end{aligned}$$

Chapter 3: Trigonometric Identities

$$= 2\sin\alpha(1-\sin^2\alpha) + (1-2\sin^2\alpha)\sin\alpha$$
$$= 2\sin\alpha - 2\sin^3\alpha + \sin\alpha - 2\sin^3\alpha$$
$$= 3\sin\alpha - 4\sin^3\alpha$$

QED

There are other ways to prove these formulas. One of them is to use the complex number method which is an important technique for solve trigonometric problems. It will be discussed in *Chapter 5*.

In addition to *(3.12)*, *(3.13)* and *(3.14)*, triple angle formulas can be written in alternative forms, as shown below.

$$\sin 3\alpha = 4 \cdot \sin(60°-\alpha) \cdot \sin\alpha \cdot \sin(60°+\alpha) \qquad (3.15)$$
$$\cos 3\alpha = 4 \cdot \cos(60°-\alpha) \cdot \cos\alpha \cdot \cos(60°+\alpha) \qquad (3.16)$$
$$\tan 3\alpha = \tan(60°-\alpha) \cdot \tan\alpha \cdot \tan(60°+\alpha) \qquad (3.17)$$

Practically, these alternative forms are used more frequently than those original forms are. Quite often, they can yield surprisingly simple solutions.

Example 3.3.2

Compute the value of $\cos 20° \cos 40° \cos 80°$.

This is a classic type of problems. A typical solution is to use the telescoping technique which will be discussed in *Chapter 4*. However, in this particular example, applying the triple angle formula offers a straightforward solution.

Solution

Noting $40° = 60° - 20°$ and $80° = 60° + 20°$ leads to

$$\cos 20° \cos 40° \cos 80°$$
$$= \cos 20° \cos(60°-20°) \cos(60°+20°)$$

$$= \frac{1}{4} \cdot \cos(3 \cdot 20°)$$
$$= \frac{1}{4} \cdot \frac{1}{2}$$
$$= \boxed{\frac{1}{8}}$$

Done.

There are several ways to prove these alternative triple angle formulas *(3.15) (3.16)*, and *(3.17)*. These proofs will be presented later when appropriate techniques are discussed.

3.4 Product-Sum and Sum-Product

An important category of identities involve transformation between product and sum of two trigonometric expressions. They are among the most frequently used formulas.

Formula *(3.18)* to *(3.21)* transform a product of two terms to a sum. They can be proved by directly expanding their right sides.

$$\sin\alpha \sin\beta = \frac{1}{2} \cdot (\cos(\alpha - \beta) - \cos(\alpha + \beta)) \qquad (3.18)$$

$$\cos\alpha \cos\beta = \frac{1}{2} \cdot (\cos(\alpha - \beta) + \cos(\alpha + \beta)) \qquad (3.19)$$

$$\sin\alpha \cos\beta = \frac{1}{2} \cdot (\sin(\alpha + \beta) + \sin(\alpha - \beta)) \qquad (3.20)$$

$$\cos\alpha \sin\beta = \frac{1}{2} \cdot (\sin(\alpha + \beta) - \sin(\alpha - \beta)) \qquad (3.21)$$

The following four formulas transform a sum of two terms to a product of two.

$$\sin\alpha + \sin\beta = 2\sin\left(\frac{\alpha + \beta}{2}\right)\cos\left(\frac{\alpha - \beta}{2}\right) \qquad (3.22)$$

Chapter 3: Trigonometric Identities

$$\cos\alpha + \cos\beta = 2\cos\left(\frac{\alpha+\beta}{2}\right)\cos\left(\frac{\alpha-\beta}{2}\right) \qquad (3.23)$$

$$\sin\alpha - \sin\beta = 2\sin\left(\frac{\alpha-\beta}{2}\right)\cos\left(\frac{\alpha+\beta}{2}\right) \qquad (3.24)$$

$$\cos\alpha - \cos\beta = -2\sin\left(\frac{\alpha+\beta}{2}\right)\sin\left(\frac{\alpha-\beta}{2}\right) \qquad (3.25)$$

One way to prove the sum-product formulas is write α and β as

$$\alpha = \frac{\alpha+\beta}{2} + \frac{\alpha-\beta}{2}, \quad \beta = \frac{\alpha+\beta}{2} - \frac{\alpha-\beta}{2}$$

Let's take *(3.22)* as an example to illustrate this technique.

Example 3.4.1

Show that $\sin\alpha + \sin\beta = 2\sin\frac{\alpha+\beta}{2}\cos\frac{\alpha-\beta}{2}$.

Proof

For any α and β, we have

$$\begin{aligned}
\sin\alpha + \sin\beta &= \sin\left(\frac{\alpha+\beta}{2} + \frac{\alpha-\beta}{2}\right) + \sin\left(\frac{\alpha+\beta}{2} - \frac{\alpha-\beta}{2}\right) \\
&= \sin\frac{\alpha+\beta}{2}\cos\frac{\alpha-\beta}{2} + \cos\frac{\alpha+\beta}{2}\sin\frac{\alpha-\beta}{2} \\
&\quad + \sin\frac{\alpha+\beta}{2}\cos\frac{\alpha-\beta}{2} - \cos\frac{\alpha+\beta}{2}\sin\frac{\alpha-\beta}{2} \\
&= 2\sin\frac{\alpha+\beta}{2}\cos\frac{\alpha+\beta}{2}
\end{aligned}$$

QED

Some trigonometric identities are similar to those in regular algebra. The two which are shown in the next example are among the most typical ones.

Example 3.4.2

Show that

$$\sin^2\alpha - \sin^2\beta = \sin(\alpha+\beta)\sin(\alpha-\beta) \qquad (3.26)$$

$$\cos^2\alpha - \cos^2\beta = -\sin(\alpha+\beta)\sin(\alpha-\beta) \qquad (3.27)$$

Proof

First applying the algebraic difference of squares formula, then sum-to-product transformation and finally the double angle formula give

$$\begin{aligned}
\sin^2\alpha - \sin^2\beta &= (\sin\alpha + \sin\beta)(\sin\alpha - \sin\beta) \\
&= \left(2\sin\frac{\alpha+\beta}{2}\cos\frac{\alpha-\beta}{2}\right)\left(2\cos\frac{\alpha+\beta}{2}\sin\frac{\alpha-\beta}{2}\right) \\
&= \left(2\sin\frac{\alpha+\beta}{2}\cos\frac{\alpha+\beta}{2}\right)\left(2\sin\frac{\alpha-\beta}{2}\cos\frac{\alpha-\beta}{2}\right) \\
&= \sin(\alpha+\beta)\sin(\alpha-\beta)
\end{aligned}$$

Similarly,

$$\begin{aligned}
\cos^2\alpha - \cos^2\beta &= (\cos\alpha + \cos\beta)(\cos\alpha - \cos\beta) \\
&= \left(2\cos\frac{\alpha+\beta}{2}\cos\frac{\alpha-\beta}{2}\right)\left(-2\sin\frac{\alpha+\beta}{2}\sin\frac{\alpha-\beta}{2}\right) \\
&= -\left(2\sin\frac{\alpha+\beta}{2}\cos\frac{\alpha+\beta}{2}\right)\left(2\sin\frac{\alpha-\beta}{2}\cos\frac{\alpha-\beta}{2}\right) \\
&= -\sin(\alpha+\beta)\sin(\alpha-\beta)
\end{aligned}$$

QED

These two formulas can also be proved by directly expanding their right sides.

Alternative Proof

Chapter 3: Trigonometric Identities

Expanding the right side and applying *(2.3)* on *page 4* lead to

$$\sin(\alpha + \beta)(\sin\alpha - \beta)$$
$$=(\sin\alpha\cos\beta + \cos\alpha\sin\beta)(\sin\alpha\cos\beta - \cos\alpha\sin\beta)$$
$$=\sin^2\alpha\cos^2\beta - \cos^2\alpha\sin^2\beta$$
$$=\sin^2\alpha(1 - \sin^2\beta) - (1 - \sin^2\alpha)\sin^2\beta$$
$$=\sin^2\alpha - \sin^2\alpha\sin^2\beta - \sin^2\beta + \sin^2\alpha\sin^2\beta$$
$$=\sin^2\alpha - \sin^2\beta$$

Meanwhile,

$$-\sin(\alpha + \beta)(\sin\alpha - \beta)$$
$$=-(\sin\alpha\cos\beta + \cos\alpha\sin\beta)(\sin\alpha\cos\beta - \cos\alpha\sin\beta)$$
$$=-\sin^2\alpha\cos^2\beta + \cos^2\alpha\sin^2\beta$$
$$=-(1 - \cos^2\alpha)\cos^2\beta + \cos^2\alpha(1 - \cos^2\beta)$$
$$=-\cos^2\beta + \cos^2\alpha\cos^2\beta + \cos^2\alpha - \cos^2\alpha\cos^2\beta$$
$$=\cos^2\alpha - \cos^2\beta$$

Done.

It is also worthy pointing out that adding the left sides of these two relations results in 0:

$$(\sin^2\alpha - \sin^2\beta) + (\cos^2\alpha - \cos^2\beta)$$
$$=(\sin^2\alpha + \cos^2\alpha) - (\sin^2\beta + \cos^2\beta)$$
$$=1 - 1$$
$$=0$$

Therefore, the right sides of these two formulas must only differ by their signs. This is indeed the case.

3.5 Practice

Practice 1

Show that $\sin\alpha - \sin\beta = 2\sin\frac{\alpha-\beta}{2}\cos\frac{\alpha+\beta}{2}$.

Practice 2

If $\sin x = \frac{2}{5}$ and x is acute, compute the values of

i) $\cos 2x$.

ii) $\cos 4x$.

iii) $\sin 2x$.

iv) $\sin 4x$.

Practice 3

Let $\alpha \in \left(\frac{3\pi}{2}, 2\pi\right)$. Simplify

$$\sqrt{\frac{1}{2} - \frac{1}{2}\sqrt{\frac{1}{2} + \frac{1}{2}\cdot\cos 2\alpha}}$$

Practice 4

Prove the following identity

$$\tan\alpha + \tan(90° - \alpha) = \frac{2}{\sin 2\alpha} \qquad (3.28)$$

Chapter 3: Trigonometric Identities

Practice 5

Prove
$$\frac{\sin(\alpha + \beta)}{\sin(\alpha - \beta)} = \frac{1 + \cot \alpha \tan \beta}{1 - \cot \alpha \tan \beta}$$

Practice 6

Prove
$$\frac{\cos(\alpha + \beta)}{\cos(\alpha - \beta)} = \frac{1 - \tan \alpha \tan \beta}{1 + \tan \alpha \tan \beta}$$

Practice 7

Solve the following equation for $0 \leq x \leq 2\pi$:
$$\sin x + \sin \frac{x}{2} = 0$$

Practice 8

Prove the following identity:
$$\sin \alpha + \sin \beta + \sin \gamma - \sin(\alpha + \beta + \gamma)$$
$$= 4 \sin \frac{\alpha + \beta}{2} \sin \frac{\beta + \gamma}{2} \sin \frac{\gamma + \alpha}{2} \qquad (3.29)$$

Practice 9

Show that if $m \tan(\theta - 30°) = n \tan(\theta + 120°)$, then
$$\cos 2\theta = \frac{m + n}{2(m - n)}$$

Practice 10

Compute the value of $(\tan 9° - \tan 27° - \tan 63° + \tan 81°)$.

Practice 11

Compute $\sin 25° \sin 35° \sin 85°$.

Practice 12

Prove $\tan 20° \tan 40° \tan 60° \tan 80° = 3$.

Practice 13

Show that $\cot 70° + 4\cos 70° = \sqrt{3}$.

Practice 14

If $\frac{1+\tan\alpha}{1-\tan\alpha} = 2018$, show that $\sec 2\alpha + \tan 2\alpha = 2018$.

Practice 15

Compute the value of $\cos 6° \cos 42° \cos 66° \cos 78°$.

Practice 16

In $\triangle ABC$, find the measurement of C if

$$3\sin A + 4\cos B = 6 \quad \text{and} \quad 4\sin B + 3\cos A = 1$$

Chapter 3: Trigonometric Identities

Practice 17

Compute the value of $(\sin^4 10° + \sin^4 50° + \sin^4 70°)$.

(Tsinghua)

Practice 18

Compute the value of $(\sin 1° \sin 2° \cdots \sin 89°)$.

Chapter 4

Trigonometric Techniques

Remembering various identities provides one pillar stone in solving trigonometry problems. Mastering proven techniques provides the other.

4.1 $a\sin\theta + b\cos\theta = \sqrt{a^2+b^2}\sin(\theta+\varphi)$

Given an expression in the form of $(a\sin\theta + b\sin\theta)$ where real numbers a and b are not both zeros, it is always possible to find an angle φ so that

$$a\sin\theta + b\cos\theta = \sqrt{a^2+b^2}\sin(\theta+\varphi) \qquad (4.1)$$

where $\tan\varphi = \frac{b}{a}$ (if $a = 0$ then $\varphi = \frac{\pi}{2}$).

Proof of this conclusion utilizes the fact that if two real number x and y satisfying $x^2 + y^2 = 1$, then it is always possible to find an angle φ so that $\sin\varphi = x$ and $\cos\varphi = y$.

Example 4.1.1

Show that transformation *(4.1)* is always possible.

Chapter 4: Trigonometric Techniques

Proof

Because a and b are not both zeros, the given expression can be rewritten in the following way:

$$a\sin\theta + b\sin\theta = \sqrt{a^2+b^2}\left(\frac{a}{\sqrt{a^2+b^2}}\sin\theta + \frac{b}{\sqrt{a^2+b^2}}\cos\theta\right)$$

Now, because

$$\left(\frac{a}{\sqrt{a^2+b^2}}\right)^2 + \left(\frac{b}{\sqrt{a^2+b^2}}\right)^2 = 1$$

therefore there must exist an angle φ such that

$$\cos\varphi = \frac{a}{\sqrt{a^2+b^2}} \quad \text{and} \quad \sin\varphi = \frac{b}{\sqrt{a^2+b^2}}$$

Setting these to the previous relation gives

$$\begin{aligned}a\sin\theta + b\cos\theta &= \sqrt{a^2+b^2}\left(\frac{a}{\sqrt{a^2+b^2}}\sin\theta + \frac{b}{\sqrt{a^2+b^2}}\cos\theta\right) \\ &= \sqrt{a^2+b^2}(\sin\theta\cos\varphi + \cos\theta\sin\varphi) \\ &= \sqrt{a^2+b^2}\sin(\theta+\varphi)\end{aligned}$$

QED

Transformation (4.1) is useful when consolidating two or more trigonometric terms into one is needed. One of its basic applications is to find maximum and minimal value of a given trigonometric expression.

Let's consider an example.

Example 4.1.2

Find the maximum and minimal values of $(\sin\alpha + \cos\alpha)$.

Solution

By *(4.1)*, it is possible to find an angle φ so that
$$\sin\alpha + \cos\alpha = \sqrt{2}\cdot\sin(\alpha+\varphi)$$
Then,
$$-1 \leq \sin(\alpha+45°) \leq 1 \implies -\sqrt{2} \leq \sin\alpha + \cos\alpha \leq \sqrt{2}$$

Done.

4.2 Multiplying and Telescoping

Many competition trigonometric problems involve a series of terms. For example:
$$\sin\alpha + \sin 2\alpha + \sin 3\alpha + \cdots + \sin n\alpha$$
$$\cos\alpha \cdot \cos 2\alpha \cdot \cos 2^2\alpha \cdots \cos 2^n\alpha$$

When these angles form an arithmetic sequence or geometric sequence, it is likely that such expression can be simplified by using the multiplying and telescoping technique.

Let's review a few examples.

Example 4.2.1

Simplify
$$\cos\frac{2\pi}{2n+1} + \cos\frac{4\pi}{2n+1} + \cdots + \cos\frac{2n\pi}{2n+1} \qquad (4.2)$$

This is a sum of an arithmetic sequence. A typical technique to simplify such an expression involves the following steps:

i) Multiplying the whole sequence by a proper trigonometric function of half of its common difference,

Chapter 4: Trigonometric Techniques

ii) Applying the product to sum transformation so that each term is break into two, with different signs,

iii) Canceling all the middle terms

Solution

Let $\theta = \frac{\pi}{2n+1}$. Then the desired sum equals

$$S = \cos 2\theta + \cos 4\theta + \cdots + \cos 2n\theta$$

The common difference here is 2θ. therefore the candidates of the to-be-multiplied term are $\sin\theta$ and $\cos\theta$. Accordingly, either formula *(3.20)* or *(3.19)* on *page 27* should be used to break each term into two.

However, *(3.19)* is not suitable here because cosine function is even which means each term will produce two terms with same sign. This will prevent us from canceling middle terms.

$$\cos\theta\cos 2\theta = \frac{1}{2}\Big(\cos(\theta - 2\theta) + \cos(\theta + 2\theta)\Big)$$
$$= \frac{1}{2}\Big(\cos\theta + \cos 3\theta\Big)$$

On the other hand, applying *(3.20)* will produce two terms with opposite signs because sine function is odd.

$$\sin\theta\cos 2\theta = \frac{1}{2}\Big(\sin(\theta - 2\theta) + \sin(\theta + 2\theta)\Big)$$
$$= \frac{1}{2}\Big(-\sin\theta + \sin 3\theta\Big)$$

Therefore, let's multiply S by $2\sin\theta$ (including a constant 2 is just for convenience). This will produce

$$2\sin\theta\cos 2\theta = \sin 3\theta - \sin\theta$$
$$2\sin\theta\cos 4\theta = \sin 5\theta - \sin 3\theta$$
$$\cdots$$

$$2\sin\theta\cos(2n\theta) = \sin(2n+1)\theta - \sin\big((2n-1)\theta\big)$$

Adding these equations together and canceling all the paired middle terms yield

$$2\sin\theta \cdot S = \sin(2n+1)\theta - \sin\theta$$

Because $\theta = \frac{\pi}{2n+1}$, we have:

$$\sin(2n+1)\theta = \sin\pi = 0$$

$$\therefore\ 2\sin\theta \cdot S = 0 - \sin\theta \implies S = \boxed{-\frac{1}{2}}$$

Done.

Let's consider another example to review the same technique again.

Example 4.2.2

Simplify

$$\sin\frac{2\pi}{2n+1} + \sin\frac{4\pi}{2n+1} + \cdots + \sin\frac{2n\pi}{2n+1} \tag{4.3}$$

Solution

Let $\theta = \frac{\pi}{2n+1}$. Therefore, the desired sum can be written as

$$S = \sin 2\theta + \sin 4\theta + \cdots + \sin 2n\theta$$

Employing the same analysis used in the previous example will lead to the conclusion that multiplying each term by $2\sin\theta$ is appropriate here because

$$2\sin\theta\sin 2\theta = \cos\theta - \cos 3\theta$$
$$2\sin\theta\sin 4\theta = \cos 3\theta - \cos 5\theta$$

Chapter 4: Trigonometric Techniques

$$\cdots$$
$$2\sin\theta\sin 2n\theta = \cos(2n-1)\theta - \cos(2n+1)\theta$$

Adding these relations results in

$$2\sin\theta \cdot S = \cos\theta - \cos(2n+1)\theta = \cos\theta - \cos\pi = \cos\theta + 1$$

Therefore, we conclude that

$$S = \frac{\cos\theta + 1}{2\sin\theta} = \boxed{\frac{\cos\frac{\pi}{2n+1} + 1}{2\sin\frac{\pi}{2n+1}}}$$

Done.

Results in the previous two examples are often used to compute the sum of a series of trigonometric terms when each individual term is difficult to compute. For example,

$$\cos\frac{2\pi}{7} + \cos\frac{4\pi}{7} + \cos\frac{6\pi}{7} = -\frac{1}{2}$$

$$\cos\frac{2\pi}{9} + \cos\frac{4\pi}{9} + \cos\frac{6\pi}{9} + \cos\frac{8\pi}{9} = -\frac{1}{2}$$

Readers are encouraged to verify these above results using WolframAlpha, Excel or other suitable calculation tools.

Likewise, a product of several trigonometric terms may also be simplified by multiplying an appropriate term. In this case, angles typically form a geometric sequence with a common ratio of 2.

Example 4.2.3

Simplify $\cos\theta \cdot \cos 2\theta \cdot \cos 2^2\theta \cdots \cos 2^n\theta$.

Solution

Let the given expression equal P. In order to create a telescoping effect, it is natural to employ the double angle formula.

Multiplying P by $\sin\theta$ gives:

$$\sin\theta \cdot P = \sin\theta \cdot \cos\theta \cdot \cos 2\theta \cdot \cos 2^2\theta \cdots \cos 2^n\theta$$

$$= \frac{1}{2} \cdot \sin 2\theta \cdot \cos 2\theta \cdot \cos 2^2\theta \cdots \cos 2^n\theta$$

$$= \frac{1}{2^2} \sin 2^2\theta \cdot \cos 2^2\theta \cdots \cos 2^n\theta$$

$$= \cdots$$

$$= \frac{1}{2^{n+1}} \sin 2^{n+1}\theta$$

$$\therefore \quad P = \boxed{\frac{\sin 2^{n+1}\theta}{2^{n+1} \cdot \sin\theta}}$$

Done.

This technique is a standard way to solve such type of problems, even though some particular expressions may be simplified using other methods. For instance, the next problem is already solved in *Example 3.3.2* on *page 26* by using the triple angle formula. It can also be tackled using the multiplying and telescoping technique.

Example 4.2.4

Compute the value of $\cos 20° \cos 40° \cos 80°$.

Alternative Solution

This is a special case of *Example 4.2.3* where $\theta = 20°$ and $n = 2$. Therefore, the result should be

$$\frac{\sin(2^{2+1} \cdot 20°)}{2^{2+1} \sin 20°} = \frac{\sin 160°}{8 \cdot \sin 20°} = \frac{\sin 20°}{8 \cdot \sin 20°} = \boxed{\frac{1}{8}}$$

This agrees with the answer obtained earlier in *Example 3.3.2*.

Chapter 4: Trigonometric Techniques

Done.

Clearly, only a series cosine terms can be tackled using this double angle formula based solution. Therefore, if the given expression is not in this form, additional transformation may be necessary.

Example 4.2.5

Compute $\sin 10° \cdot \sin 50° \cdot \sin 70°$.

Solution

Because $\sin \alpha = \cos(90° - \alpha)$, we find

$$\sin 10° \cdot \sin 50° \cdot \sin 70° = \cos 80° \cdot \cos 40° \cdot \cos 20°$$

This is the same as the problem in the previous example. Therefore, the result is also $\boxed{\dfrac{1}{8}}$.

Done.

4.3 Reverse Construction

In most cases, a trigonometric problem is tackled by converting a trigonometric function to its equivalent numerical value. For example, $\sin 30° \to \frac{1}{2}$. However, in some cases, it is useful to perform in the reverse way, i.e., converting a number back to an equivalent trigonometric expression, e.g. $\frac{1}{2} \to \sin 30°$. The objective of such reverse construction is to allow application of certain trigonometric transformations which may help derive the final result easier.

In fact, some previous examples have already employed this technique. For instance, the transformation discussed in *Section 4.1*

$$(a\sin\theta + b\cos\theta) \Rightarrow \sqrt{a^2 + b^2}\sin(\theta + \varphi)$$

Chapter 4: Trigonometric Techniques

is such an example.

Let's consider another example. This is to prove one of the alternative triple angle formulas.

Example 4.3.1

Given $\sin 3\theta = 3\sin\theta - 4\sin^2\theta$, prove
$$\sin 3\theta = 4\sin\theta \sin(60° + \theta)\sin(60° - \theta)$$

Proof

The interesting thing about this formula is the seemingly abrupt appearance of 60°. Meanwhile, the product of $\sin(60° + \theta)$ and $\sin(60° - \theta)$ appears to be related to the difference of square formula *(3.26)* on *page 29*. Finally, given $\sin^2 60° = \frac{3}{4}$ which apparently links to the coefficients of 3 and 4, we find a perfect case of applying the reverse construction technique.

$$\begin{aligned}
\sin 3\theta &= 3\sin\theta - 4\sin^3\theta \\
&= 4\sin\theta \cdot \left(\frac{3}{4} - \sin^2\theta\right) \\
&= 4\sin\theta \cdot \left(\sin^2 60° - \sin^2\theta\right) \\
&= 4\sin\theta \sin(60° + \theta)\sin(60° - \theta)
\end{aligned}$$

QED

Another frequently seen case of reverse construction utilizes the fact of $\tan 45° = 1$. Combining the sum of tangent formula gives

$$\frac{1 \pm \tan\theta}{1 \mp \tan\theta} = \frac{\tan 45° \pm \tan\theta}{1 \mp \tan\theta \tan 45°} = \tan(45° \pm \theta) \qquad (4.4)$$

This transformation will be used later.

Chapter 4: Trigonometric Techniques

4.4 Symmetry and Pairing

Some trigonometric expressions are symmetric in the sense that the sum of paired terms is a constant which often equals a special angle. For example:

$$(1+\tan 1°)(1+\tan 2°)\cdots(1+\tan 43°)(1+\tan 44°)$$

This resembles the feature of arithmetic sequence in regular algebra. The sum formula of an arithmetic sequence can be obtained by pairing symmetric terms whose sum is a constant. A similar technique can also be employed to tackle such trigonometric expressions.

Example 4.4.1

Compute the value of

$$(1+\tan 1°)(1+\tan 2°)\cdots(1+\tan 43°)(1+\tan 44°)$$

Solution

Noting that this expression contains 22 pairs whose corresponding angles sum to a special angle 45°. Let's first try to evaluate the following expression:

$$(1+\tan\alpha)(1+\tan(45°-\alpha)) \tag{4.5}$$

It turns out that the sum of such a pair equals a constant because

$$\begin{aligned}
&(1+\tan\alpha)(1+\tan(45°-\alpha)) \\
&= (1+\tan\alpha)\left(1+\frac{\tan 45°-\tan\alpha}{1+\tan 45°\tan\alpha}\right) \\
&= (1+\tan\alpha)\left(1+\frac{1-\tan\alpha}{1+\tan\alpha}\right) \\
&= 2
\end{aligned}$$

Hence, the original expression equals $\boxed{2^{22}}$.

Done.

Expression *(4.5)* can also be evaluated using *(3.5)* on *page 22* because their sum is a special angle.

$$\begin{aligned}
&(1+\tan\alpha)(1+\tan(45°-\alpha)) \\
=\ &1+\tan\alpha\tan(45°-\alpha)+\left(\tan\alpha+\tan(45°-\alpha)\right) \\
=\ &1+\tan\alpha\tan(45°-\alpha)+\tan 45°(1-\tan\alpha\tan(45°-\alpha)) \\
=\ &2
\end{aligned}$$

4.5 Mathematical Induction

The vast majority of competition trigonometric problems can be solved by applying basic identities and employing appropriate techniques. Mathematical induction usually is not the method of choice. However, when the given expression is so simple that makes applying trigonometric transformation less obvious, induction may still be a useful choice.

Let's consider one example here.

Example 4.5.1

Let x be a real number and n be a positive integer. Show that

$$n\mid\sin x\mid\ \geq\ \mid\sin nx\mid$$

Proof

The claim obviously holds when $n=1$. Assume it also holds when $n=k$, i.e. $k\mid\sin x\mid\ \geq\ \mid\sin kx\mid$. Then when $n=k+1$, we have

$$\mid\sin(k+1)x\mid$$

$$= |\sin kx \cos x + \cos kx \sin x|$$
$$\leq |\sin kx \cos x| + |\cos kx \sin x|$$
$$= |\sin kx| \cdot |\cos x| + |\cos kx| \cdot |\sin x|$$
$$\leq |\sin kx| + |\sin x|$$
$$\leq k|\sin x| + |\sin x|$$
$$= (k+1)|\sin x|$$

Therefore, by the principle of mathematical induction, the claim always holds.

QED

4.6 Half-Tangent Substitution

The half-tangent substitution is widely used in calculus. Its objective is to transform an expression containing several different trigonometric functions of θ to one which contains just $\tan \frac{\theta}{2}$. The transformation is usually done by letting $t = \tan \frac{\theta}{2}$. Then, $\sin \theta$, $\cos \theta$ and $\tan \theta$ all can be written with respect to t:

$$\sin \theta = \frac{2t}{1+t^2} \tag{4.6}$$

$$\cos \theta = \frac{1-t^2}{1+t^2} \tag{4.7}$$

$$\tan \theta = \frac{2t}{1-t^2} \tag{4.8}$$

Identity *(4.8)* is a direct result of the double angle formula *(3.8)* on *page 22*. The other two can be proved in several different ways. One of them is a beautiful application of the reverse construction technique discussed earlier.

Example 4.6.1

Prove the following two identities:

$$\sin\theta = \frac{2\tan\frac{\theta}{2}}{1+\tan^2\frac{\theta}{2}}, \qquad \cos\theta = \frac{1-\tan^2\frac{\theta}{2}}{1+\tan^2\frac{\theta}{2}}$$

Proof

First, rewrite $\sin\theta$ and $\cos\theta$ as

$$\sin\theta = \frac{\sin\theta}{\sin^2\frac{\theta}{2}+\cos^2\frac{\theta}{2}} = \frac{2\sin\frac{\theta}{2}\cos\frac{\theta}{2}}{\sin^2\frac{\theta}{2}+\cos^2\frac{\theta}{2}}$$

$$\cos\theta = \frac{\cos\theta}{\sin^2\frac{\theta}{2}+\cos^2\frac{\theta}{2}} = \frac{\cos^2\frac{\theta}{2}-\sin^2\frac{\theta}{2}}{\sin^2\frac{\theta}{2}+\cos^2\frac{\theta}{2}}$$

Then, dividing both the denominator and the numerator by $\cos^2\frac{\theta}{2}$ leads to the desired results immediately.

QED

4.7 Practice

Practice 1

Show that

$$\cos(\alpha)+\cos(\alpha+\beta)+\cdots+\cos(\alpha+n\beta) = \frac{\sin\frac{n+1}{2}\beta \cos\left(\alpha+\frac{n}{2}\beta\right)}{\sin\frac{\beta}{2}}$$

Practice 2

Show that

$$\cos\alpha + \cos 2\alpha + \cos 3\alpha + \cdots + \cos n\alpha = \frac{\sin\frac{n}{2}\alpha \cos\frac{n+1}{2}\alpha}{\sin\frac{\alpha}{2}}$$

Practice 3

Show that for any natural number n and real number $x \neq \frac{m\pi}{2^k}$ (where m in an integer), the following relation always holds:

$$\frac{1}{\sin 2x} + \frac{1}{\sin 4x} + \cdots + \frac{1}{\sin 2^n x} = \cot x - \cot 2^n x$$

Practice 4

Show that for any positive integer:

$$\tan x \tan 2x + \tan 2x \tan 3x + \cdots + \tan(n-1)x \tan nx = \frac{\tan nx}{\tan x} - n$$

Practice 5

Show that

$$\tan x + 2\tan 2x + 2^2 \tan 2^2 x + \cdots + 2^n \tan 2^n x = \cot x - 2^{n+1} \cot 2^{n+1} x$$

Practice 6

Show that

$$\cos\frac{\pi}{7} + \cos\frac{3\pi}{7} + \cos\frac{5\pi}{7} = \frac{1}{2}$$

Practice 7

Compute the value of $(\cos\frac{\pi}{7} - \cos\frac{2\pi}{7} + \cos\frac{3\pi}{7})$.

Practice 8

Compute the value of $(\cos\frac{\pi}{9} + \cos\frac{3\pi}{9} + \cos\frac{5\pi}{9} + \cos\frac{7\pi}{9})$.

Practice 9

Show that

$$\frac{1}{\sin 1° \sin 2°} + \frac{1}{\sin 2° \sin 3°} + \cdots + \frac{1}{\sin 89° \sin 90°} = \cos 1° \csc^2 1°$$

Practice 10

Compute the value of $(\sqrt{3}\tan 18° + \tan 18° \tan 12° + \sqrt{3}\tan 12°)$.

Practice 11

Compute

$$\cos\frac{\pi}{2n+1} \cdot \cos\frac{2\pi}{2n+1} \cdots \cos\frac{n\pi}{2n+1}$$

Practice 12

Prove that $\cos 1°$ is irrational.

Chapter 4: Trigonometric Techniques

Practice 13

Find the minimal value of

$$|\sin x + \cos x + \tan x + \cot x + \sec x + \csc x|$$

where x is a real number.

(Putnam)

Chapter 5

The Complex Number Method

Trigonometry is closely related to complex number. Many complex number problems can be solved using trigonometry. At the same time, complex number also provides a powerful tool to solve certain trigonometry problems.

5.1 Trigonometry to Complex Number

There are several ways which complex number can be used to solve trigonometry problems. Most of them just involve basic well-known relations such as the De Moivre law. They will be illustrated using examples in the next section. Having said this, there is one relatively less known but useful trigonometry to complex transformation. It is explained below.

Let $z = \cos\theta + i\sin\theta$, then $\bar{z} = \frac{1}{z} = \cos\theta - i\sin\theta$ and

$$\cos\theta = \frac{1}{2}\cdot\left(z + \frac{1}{z}\right) \quad , \quad \sin\theta = \frac{1}{2i}\cdot\left(z - \frac{1}{z}\right) \qquad (5.1)$$

Chapter 5: The Complex Number Method

Meanwhile, the De Moivre's theorem states that $z^n = (\cos\theta + i\sin\theta)^n = \cos n\theta + i \sin n\theta$. Therefore,

$$\cos n\theta = \frac{1}{2} \cdot \left(z^n + \frac{1}{z^n}\right), \quad \sin n\theta = \frac{1}{2i} \cdot \left(z^n - \frac{1}{z^n}\right) \qquad (5.2)$$

Formulas (5.1) and (5.2) can be used to transform a trigonometric expression of θ to an algebraic expression with respect to z. This means that some trigonometric expressions can be dealt with using various algebraic techniques after such transformation. While this method may require some computational efforts, it can offer meaningful value because most students are more confident and skilled in handling algebraic expressions than in tackling trigonometric ones.

5.2 Examples

Let's start by revisiting the triple angle formula. It serves as a good example to show that many trigonometric identities can be proved by the complex number method. It is worth pointing out that the complex number method can often prove paired sine and cosine identities at once.

Example 5.2.1

Prove $\sin 3\theta = 3\sin\theta - 4\sin^3\theta$ and $\cos 3\theta = 4\cos^3\theta - 3\cos\theta$.

Proof

Let $z = \cos\theta + i\sin\theta$. Then by the De Moivre law, we have

$$z^3 = \cos 3\theta + i \sin 3\theta \qquad (5.3)$$

Meanwhile, z^3 can also be evaluated using binomial expansion:

$$\begin{aligned} z^3 &= (\cos\theta + i\sin\theta)^3 \\ &= \cos^3\theta + 3\cos^2\theta\sin\theta(i) + 3\cos\theta\sin^2\theta(i^2) + \sin^3\theta(i^3) \\ &= (\cos^3\theta - 3\cos\theta\sin^2\theta) + i(3\cos^2\theta\sin\theta - \sin^3\theta) \end{aligned}$$

Chapter 5: The Complex Number Method

$$= (\cos^3\theta - 3\cos\theta(1-\cos^2\theta)) + i(3(1-\sin^2\theta)\sin\theta - \sin^3\theta)$$
$$= (4\cos^3\theta - 3\cos\theta) + i(3\sin\theta - 4\sin^3\theta)$$

This result has to agree with (5.3). Matching their real and imaginary parts, respectively, leads to the desired result immediately.

QED

The next problem is a revisit of an earlier example (4.2.1) on page 37. Instead of applying trigonometric transformation, it employs regular algebraic simplification techniques.

Example 5.2.2

Compute

$$\cos\frac{2\pi}{2n+1} + \cos\frac{4\pi}{2n+1} + \cdots + \cos\frac{2n\pi}{2n+1}$$

Solution

Let $\theta = \frac{2\pi}{2n+1}$ and $z = \cos\theta + i\sin\theta$. Then $z^{2n+1} = 1$.

$$\therefore \cos\frac{2\pi}{2n+1} + \cos\frac{4\pi}{2n+1} + \cdots + \cos\frac{2n\pi}{2n+1}$$
$$= \frac{1}{2}\cdot\left(z+\frac{1}{z}\right) + \frac{1}{2}\left(z^2+\frac{1}{z^2}\right) + \cdots + \frac{1}{2}\left(z^n+\frac{1}{z^n}\right)$$
$$= \frac{1}{2}\cdot\left((z+z^2+\cdots+z^n) + \left(\frac{1}{z}+\frac{1}{z^2}+\cdots+\frac{1}{z^n}\right)\right)$$
$$= \frac{1}{2}\cdot\left(z\cdot\frac{1-z^n}{1-z} + \frac{1}{z}\cdot\frac{1-\frac{1}{z^n}}{1-\frac{1}{z}}\right)$$
$$= \frac{1}{2}\cdot\left(\frac{z-z^{n+1}}{1-z} + \frac{1-\frac{1}{z^n}}{z-1}\right)$$
$$= \frac{1}{2}\cdot\left(\frac{z-z^{n+1}}{1-z} + \frac{1-\frac{z^{2n+1}}{z^n}}{z-1}\right)$$

Chapter 5: The Complex Number Method

$$= \frac{1}{2} \cdot \left(\frac{z - z^{n+1}}{1-z} + \frac{1 - z^{n+1}}{z-1} \right)$$

$$= \boxed{-\frac{1}{2}}$$

Done.

The result in *Example 5.2.2* is a constant. If this is not the case, then it will be necessary to convert an algebraic expression of z back to a trigonometric expression of θ at the end.

Example 5.2.3

Simplify $(\cos\alpha + \cos 2\alpha + \cos 3\alpha + \cdots + \cos n\alpha)$.

This problem can be solved by the telescoping technique. Meanwhile, it is also an excellent example to demonstrate the conversion from an expression of z back to one with respect to θ because its result is not a constant.

Proof

Let $z = \cos\alpha + i\sin\alpha$, then

$$\cos\alpha + \cos 2\alpha + \cos 3\alpha + \cdots + \cos n\alpha$$

$$= \frac{1}{2} \cdot \left(\left(z + \frac{1}{z}\right) + \left(z^2 + \frac{1}{z^2}\right) + \left(z^3 + \frac{1}{z^3}\right) + \cdots + \left(z^n + \frac{1}{z^n}\right) \right)$$

$$= \frac{1}{2} \cdot \left((z + z^2 + z^3 + \cdots + z^n) + \left(\frac{1}{z} + \frac{1}{z^2} + \frac{1}{z^3} + \cdots + \frac{1}{z^n}\right) \right)$$

$$= \frac{1}{2} \cdot \left(z \cdot \frac{1-z^n}{1-z} + \frac{1}{z} \cdot \frac{1 - \frac{1}{z^n}}{1 - \frac{1}{z}} \right)$$

$$= \frac{1}{2} \cdot \frac{1}{z^n} \cdot \frac{1}{1-z} \cdot (z^{n+1} - z^{2n+1} - z^n + 1)$$

$$= \frac{1}{2} \cdot \frac{1}{z^n} \cdot \frac{1}{1-z} \cdot (1 - z^n)(z^{n+1} + 1)$$

Chapter 5: The Complex Number Method

Now, we need to transform the expression to a product of $(z^k \pm \frac{1}{z^k})$ so that *(5.2)* can be used to convert the result back to a trigonometric expression. Note that that target is "symmetric" with respect to z^k and $\frac{1}{z^k}$. Therefore, when the exponents of the two terms are different, the average of their exponents can be used.

$$\frac{1}{1-z} = \frac{1}{z^{-\frac{1}{2}} - z^{\frac{1}{2}}} \cdot \frac{1}{z^{\frac{1}{2}}}$$

$$(1 - z^n) = \left(z^{-\frac{n}{2}} - z^{\frac{n}{2}}\right) \cdot z^{\frac{n}{2}}$$

$$z^{n+1} + 1 = \left(z^{\frac{n+1}{2}} + z^{-\frac{n+1}{2}}\right) \cdot z^{\frac{n+1}{2}}$$

Setting these back to the original expression yields

$$\frac{1}{2} \cdot \frac{1}{z^n} \cdot \frac{1}{z^{-\frac{1}{2}} - z^{\frac{1}{2}}} \cdot \frac{1}{z^{\frac{1}{2}}} \cdot \left(z^{-\frac{n}{2}} - z^{\frac{n}{2}}\right) \cdot z^{\frac{n}{2}} \cdot \left(z^{\frac{n+1}{2}} + z^{-\frac{n+1}{2}}\right) \cdot z^{\frac{n+1}{2}}$$

$$= \frac{1}{2} \cdot \frac{1}{z^{-\frac{1}{2}} - z^{\frac{1}{2}}} \cdot \left(z^{-\frac{n}{2}} - z^{\frac{n}{2}}\right) \cdot \left(z^{\frac{n+1}{2}} + z^{-\frac{n+1}{2}}\right)$$

$$= \frac{1}{2} \cdot \frac{1}{-2i \sin \frac{\alpha}{2}} \cdot \left(-2i \sin \frac{n\alpha}{2}\right) \cdot \left(2 \cos \frac{(n+1)\alpha}{2}\right)$$

$$= \frac{1}{\sin \frac{\alpha}{2}} \cdot \left(\sin \frac{n\alpha}{2}\right) \cdot \left(\cos \frac{(n+1)\alpha}{2}\right)$$

$$= \boxed{\frac{\sin \frac{n\alpha}{2} \cos \frac{(n+1)\alpha}{2}}{\sin \frac{\alpha}{2}}}$$

QED

Complex number has a rich set of tools and theorems. Hence, the complex number method is a versatile technique. Let's consider another example.

Chapter 5: The Complex Number Method

Example 5.2.4

Show that

$$\sin\alpha + \sin(\alpha + 120°) + \sin(\alpha - 120°) = 0$$

$$\cos\alpha + \cos(\alpha + 120°) + \cos(\alpha - 120°) = 0$$

Thee two identities can be proved by directly expanding their left sides. Here, a vector based explanation is presented to show the intrinsic meanings of these two relations.

Proof

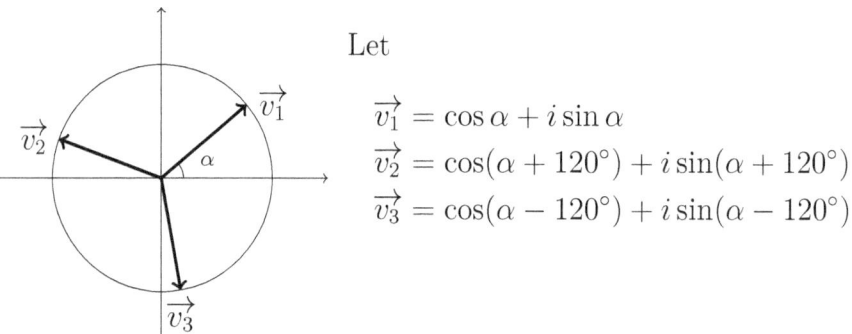

Let

$$\vec{v_1} = \cos\alpha + i\sin\alpha$$
$$\vec{v_2} = \cos(\alpha + 120°) + i\sin(\alpha + 120°)$$
$$\vec{v_3} = \cos(\alpha - 120°) + i\sin(\alpha - 120°)$$

Then, by the basic property of vectors, we can assert that vectors $\vec{v_1}$, $\vec{v_2}$ and $\vec{v_3}$ can form an enclosed equilateral triangle which means that[1]

$$\vec{v_1} + \vec{v_2} + \vec{v_3} = 0$$

This means both their real and imaginary parts must equal zero.

QED

The vector based explanation can also be used to easily show the following extension of *Example 5.2.4* by considering rotating $\vec{v_1}$, $\vec{v_2}$,

---

[1] This can also be explained using Lami's theorem which is discussed in the book *Geometry Technique* written by the same author.

and $\vec{v_3}$.

Example 5.2.5

Let k be a positive integer, then

$$\sin k\alpha + \sin k(\alpha + 120°) + \sin k(\alpha - 120°) = 0$$

$$\cos k\alpha + \cos k(\alpha + 120°) + \cos k(\alpha - 120°) = 0$$

We will skip the proof here and prove an enhanced version of this claim in the practice.

5.3 Practice

Practice 1

Prove the following formula using the complex number method:

$$\sin(\alpha + \beta) = \sin\alpha \cos\beta + \cos\alpha \sin\beta$$

$$\cos(\alpha + \beta) = \cos\alpha \cos\beta - \sin\alpha \sin\beta$$

Practice 2

Solve the equation $\cos^2 x + \cos^2 2x + \cos^2 3x = 1$ in $(0, 2\pi)$.
(1962 IMO)

Chapter 5: The Complex Number Method

Practice 3

Prove the following identities
$$\sin(3\theta) = 3\sin\theta - 4\sin^3\theta$$
$$\cos(3\theta) = 4\cos^3\theta - 3\cos\theta$$

Practice 4

Solve the equation $\cos\theta + \cos 2\theta + \cos 3\theta = \sin\theta + \sin 2\theta + \sin 3\theta$.

Practice 5

If $\sin A + \sin B + \sin C = 0 = \cos A + \cos B + \cos C = 0$, explain for any positive integer n, it must hold that

$$\sin nA + \sin nB + \sin nC = \cos nA + \cos nB + \cos nC = 0$$

Practice 6

Prove: $\cos 7x + 7\cos 5x + 21\cos 3x + 35\cos x = 64\cos^7 x$.

Practice 7

Let $A(x_1, y_1)$, $B(x_2, y_2)$, and $C(x_3, y_3)$ be three points on the unit circle, and
$$x_1 + x_2 + x_3 = y_1 + y_2 + y_3 = 0$$
Prove
$$x_1^2 + x_2^2 + x_3^2 = y_1^2 + y_2^2 + y_3^2 = \frac{3}{2}$$

Practice 8

Compute the values of

$$S = C_n^1 \sin\theta + C_n^2 \sin 2\theta + \cdots + C_n^n \sin n\theta$$

and

$$C = C_n^1 \cos\theta + C_n^2 \cos 2\theta + \cdots + C_n^n \cos n\theta$$

Chapter 5: The Complex Number Method

Chapter 6

Trigonometry in Triangle

Trigonometry is a powerful tool to solve some geometry problems, especially when triangles ae involved. Consequently, it is important to study the trigonometric relations in the context of a triangle. Some topics are covered in the books *Geometry Technique* and *Geometry Theorem*. While they are included here for completeness, the focus of this chapter is triangular trigonometric identities and inequalities.

For the sake of conciseness, A, B, C are used to denote the three interior angles and a, b, c represent lengths of the three sides opposite to A, B, C, respectively. Additionally, R is $\triangle ABC$'s circumradius and r is its inradius.

6.1 Law of Sines, Cosines and Tangents

The law of sines and the law of cosines are the two most basic triangular trigonometric relations. They state that

$$\frac{a}{\sin A} = \frac{b}{\sin B} = \frac{c}{\sin C} = 2R \tag{6.1}$$

Chapter 6: Trigonometry in Triangle

$$\begin{cases} a^2 = b^2 + c^2 - 2bc\cos A \\ b^2 = c^2 + a^2 - 2ca\cos B \\ c^2 = a^2 + b^2 - 2ab\cos C \end{cases} \quad (6.2)$$

A direct derivation of the law of sines *(6.1)* is that the ratio of two sides equals the ratio of the sine values of two corresponding angles:

$$\frac{a}{b} = \frac{\sin A}{\sin B}, \quad \frac{b}{c} = \frac{\sin B}{\sin C}, \quad \frac{c}{a} = \frac{\sin C}{\sin A} \quad (6.3)$$

Meanwhile, law of cosines *(6.2)* is often rewritten as

$$\cos A = \frac{b^2 + c^2 - a^2}{2bc} \quad (6.4)$$

These two laws can be used together with various trigonometric identities to produce many interesting result. Let's review one of them here.

Example 6.1.1

In $\triangle ABC$, it must hold that

$$\sin\frac{A}{2} = \sqrt{\frac{(p-b)(p-c)}{bc}} \quad (6.5)$$

where $p = \frac{1}{2}(a+b+c)$.

Proof

By the half angle formula *(3.9)* on *page 23*, we have

$$\sin^2\frac{A}{2} = \frac{1-\cos A}{2}$$

Replacing $\cos A$ with *(6.4)* above leads to

$$\sin^2\frac{A}{2} = \frac{1-\cos A}{2}$$

$$= \frac{1}{2} \cdot \left(1 - \frac{b^2+c^2-a^2}{2bc}\right)$$
$$= \frac{1}{2} \cdot \frac{1}{2bc} \cdot \left(2bc - (b^2+c^2-a^2)\right)$$
$$= \frac{1}{4bc} \cdot \left(a^2 - (b^2+c^2-2bc)\right)$$
$$= \frac{1}{4bc} \cdot \left(a^2 - (b-c)^2\right)$$
$$= \frac{1}{4bc} \cdot (a+b-c)(a-b+c)$$
$$= \frac{1}{4bc} \cdot ((a+b+c) - 2c)((a+b+c) - 2b)$$
$$= \frac{1}{4bc} \cdot (2p-2c)(2p-2b)$$
$$= \frac{(p-b)(p-c)}{bc}$$
$$\therefore \quad \sin \frac{A}{2} = \sqrt{\frac{(p-b)(p-c)}{bc}}$$

QED

While relatively less well known than its two cousins, the law of tangents can come handy when sums and differences of two sides or two angles are involved. This law states that

$$\frac{\tan \frac{A-B}{2}}{\tan \frac{A+B}{2}} = \frac{\frac{a-b}{2}}{\frac{a+b}{2}} \qquad (6.6)$$

This relation can be proved by first applying the law of sines and then the sum-product transformation.

$$\frac{\frac{a-b}{2}}{\frac{a+b}{2}} = \frac{a-b}{a+b}$$
$$= \frac{2R\sin A - 2R\sin B}{2R\sin A + 2R\sin B}$$
$$= \frac{\sin A - \sin B}{\sin A + \sin B}$$

Chapter 6: Trigonometry in Triangle

$$= \frac{2\sin\frac{A-B}{2}\cos\frac{A+B}{2}}{2\sin\frac{A+B}{2}\cos\frac{A-B}{2}}$$

$$= \frac{\tan\frac{A-B}{2}}{\tan\frac{A+B}{2}}$$

6.2 Areas, Circumradius and Inradius

Applying the law of sines to the formula $S_{\triangle ABC} = \frac{1}{2} \cdot a \cdot b \cdot \sin C$ yields the following formulas:

$$S_{\triangle ABC} = 2R^2 \sin A \sin B \sin C = \frac{abc}{4R} \qquad (6.7)$$

In addition to *(6.7)*, there are many R and r related identities. Most of them involve trigonometry and often can be proved using the area method.

Let's look at one example.

Example 6.2.1

In $\triangle ABC$, show that

$$r = 4R \sin\frac{A}{2} \sin\frac{B}{2} \sin\frac{C}{2} \qquad (6.8)$$

Proof

By *(6.5)* on *page 62*, we have

$$\sin\frac{A}{2} = \sqrt{\frac{(p-b)(p-c)}{bc}}$$

where $p = \frac{1}{2}(a+b+c)$. Similarly, we have

$$\sin\frac{B}{2} = \sqrt{\frac{(p-c)(p-a)}{bc}} \quad \text{and} \quad \sin\frac{C}{2} = \sqrt{\frac{(p-a)(p-b)}{ab}}$$

Multiplying these three relations together yields
$$\sin\frac{A}{2}\sin\frac{B}{2}\sin\frac{C}{2} = \frac{(p-a)(p-b)(p-c)}{abc}$$
By Heron's formula, we find
$$(p-a)(p-b)(p-c) = \frac{S^2_{\triangle ABC}}{p}$$
$$\therefore \quad 4R\sin\frac{A}{2}\sin\frac{B}{2}\sin\frac{C}{2}$$
$$= 4R \cdot \frac{1}{abc} \cdot \frac{S^2_{\triangle ABC}}{p}$$
$$= \frac{1}{S_{\triangle ABC}} \cdot \frac{S^2_{\triangle ABC}}{p}$$
$$= \frac{S_{\triangle ABC}}{p}$$
$$= r$$

QED

The proof of the preceding example uses the formula $S_{\triangle ABC} = rp$. Proof of this formula is presented in the book *Geometry Techniques*.

6.3 Trigonometric Ceva's Theorem

Ceva's theorem is one of the most used geometry theorems in math competitions. Comparing to its regular form, the trigonometric form of the Ceva's theorem is relatively less known.

Let O be a point not locating on the sides of $\triangle ABC$, then the following relation always holds
$$\frac{\sin \angle CAO}{\sin \angle BAO} \cdot \frac{\sin \angle ABO}{\sin \angle CBO} \cdot \frac{\sin \angle BCO}{\sin \angle ACO} = 1 \qquad (6.9)$$

Chapter 6: Trigonometry in Triangle

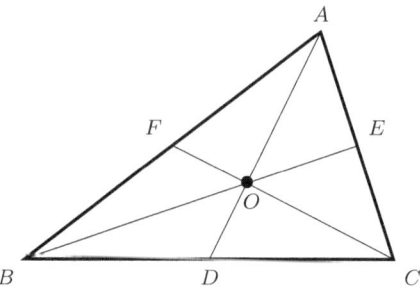

Relation (6.9) can be proved in a similar way as the regular form by employing the area method.

$$
\begin{aligned}
1 &= \frac{S_{\triangle CAO}}{S_{\triangle BAO}} \cdot \frac{S_{\triangle ABO}}{S_{\triangle CBO}} \cdot \frac{S_{\triangle BCO}}{S_{\triangle ACO}} \\
&= \frac{AC \cdot AO \sin \angle CAO/2}{AB \cdot AO \sin \angle BAO/2} \cdot \frac{AB \cdot BO \sin \angle ABO/2}{CB \cdot BO \sin \angle CBO/2} \\
&\quad \cdot \frac{BC \cdot CO \sin \angle BAO/2}{AC \cdot CO \sin \angle ACO/2} \\
&= \frac{\sin \angle CAO}{\sin \angle BAO} \cdot \frac{\sin \angle ABO}{\sin \angle CBO} \cdot \frac{\sin \angle BCO}{\sin \angle ACO}
\end{aligned}
$$

6.4 Triangular Trigonometric Identities

When a trigonometric identity involves three positive angles whose sum equals 180°, then this identity becomes a triangular identity. This is because these three angles can be the three interior angles of a triangle.

The vast majority of triangular identities are special cases of corresponding general trigonometric relations. As such, they can be proved by setting $A + B + C = 180°$ in the corresponding general trigonometric identities. Alternatively, it is also possible to prove them directly using appropriate transformations and techniques in a similar manner as to prove general trigonometric identities.

Let's consider a couple of examples.

Example 6.4.1

If $\triangle ABC$ is not a right triangle, show

$$\tan A + \tan B + \tan C = \tan A \tan B \tan C \qquad (6.10)$$

Because a triangular trigonometric identity typically involves three angles, its proof usually proceeds by grouping A and B together first, and then setting $C = (180° - (A+B))$.

Proof

Given $\triangle ABC$ is not right, $\tan A$, $\tan B$, and $\tan C$ are all well-defined. Applying *(3.5)* on *page 22*, leads to

$$\begin{aligned}
& \tan A + \tan B + \tan C \\
=\ & \tan(A+B)(1 - \tan A \tan B) + \tan C \\
=\ & \tan(180° - C)(1 - \tan A \tan B) + \tan C \\
=\ & -\tan C(1 - \tan A \tan B) + \tan C \\
=\ & \tan A \tan B \tan C
\end{aligned}$$

$$\text{QED}$$

Example 6.4.2

In $\triangle ABC$, show that

$$\sin A + \sin B + \sin C = 4\cos\frac{A}{2}\cos\frac{B}{2}\cos\frac{C}{2} \qquad (6.11)$$

A Direct Proof

Note that $A + B + C = 180° \implies \sin C = \sin(A+B)$, we have

$$\sin A + \sin B + \sin C$$

Chapter 6: Trigonometry in Triangle

$$\begin{aligned}
&= \sin A + \sin B + \sin(A+B) \\
&= 2\sin\frac{A+B}{2}\cos\frac{A-B}{2} + 2\sin\frac{A+B}{2}\cos\frac{A+B}{2} \\
&= 2\sin\frac{A+B}{2}\left(\cos\frac{A-B}{2} + \cos\frac{A+B}{2}\right) \\
&= 2\sin\left(90° - \frac{C}{2}\right)\left(2\cos\frac{A}{2}\cos\frac{B}{2}\right) \\
&= 4\cos\frac{A}{2}\cos\frac{B}{2}\cos\frac{C}{2}
\end{aligned}$$

Done.

An Indirect Proof

The to be proved identity is a special case of *(8.5)* on *page 99*.

$$\sin\alpha + \sin\beta + \sin\gamma - \sin(\alpha+\beta+\gamma) = 4\sin\frac{\alpha+\beta}{2}\sin\frac{\beta+\gamma}{2}\sin\frac{\gamma+\alpha}{2}$$

Setting $\alpha = A$, $\beta = B$, $\gamma = C$ where $A + B + C = 180°$ yields

$$\begin{aligned}
&\sin A + \sin B + \sin C - \sin(A+B+C) \\
&= 4\sin\frac{A+B}{2}\sin\frac{B+C}{2}\sin\frac{C+A}{2} \\
\Leftrightarrow\quad &\sin A + \sin B + \sin C - \sin 180° \\
&= 4\sin\left(90° - \frac{C}{2}\right)\sin\left(90° - \frac{A}{2}\right)\sin\left(90° - \frac{B}{2}\right) \\
\Leftrightarrow\quad &\sin A + \sin B + \sin C = 4\cos\frac{C}{2}\cos\frac{A}{2}\cos\frac{B}{2} \\
\Leftrightarrow\quad &\sin A + \sin B + \sin C = 4\cos\frac{A}{2}\cos\frac{B}{2}\cos\frac{C}{2}
\end{aligned}$$

Done.

There exists a large quantity of triangular identities. Usually, it is unnecessary to remember all of them, especially at the beginner to

Chapter 6: Trigonometry in Triangle

intermediate level. However, becoming familiar with these relations is helpful to solve certain problems at the advanced level when the ability of making reasonable conjectures plays an important and differentiating role in math competitions.

6.5 Triangular Trigonometric Inequalities

Similar to triangular trigonometric identities, many inequalities involving three positive angles adding up 180° also exist. For example,
$$\tan A \cdot \tan B \cdot \tan C \geq 3\sqrt{3}$$

A typical approach to prove such inequalities is to find its corresponding triangular identities and then to apply the AM-GM inequality.

The AM-GM inequality states that given n non-negative real numbers $x_1, x_2, \cdots, x_n$, the following relation always holds

$$\frac{x_1 + x_2 + \cdots + x_n}{n} \geq \sqrt[n]{x_1 \cdot x_2 \cdots x_n} \qquad (6.12)$$

which is equivalent to asserting

$$x_1 + x_2 + \cdots + x_n \geq n \cdot \sqrt[n]{x_1 \cdot x_2 \cdots x_n} \qquad (6.13)$$

Let's illustrate these points using the following example:

Example 6.5.1

Given $\triangle ABC$, show that
$$\tan A \cdot \tan B \cdot \tan C \geq 3\sqrt{3}$$

Chapter 6: Trigonometry in Triangle

Proof

The corresponding identity relating to the to-be-proved claim is *(6.10)* on *page 67*:

$$\tan A + \tan B + \tan C = \tan A \cdot \tan B \cdot \tan C$$

Applying the AM-GM inequity, *(6.13)*, leads to

$$\tan A \cdot \tan B \cdot \tan C$$
$$= \tan A + \tan B + \tan C$$
$$\geq 3 \cdot \sqrt[3]{\tan A \cdot \tan B \cdot \tan C}$$
$$\implies (\tan A \cdot \tan B \cdot \tan C)^3 \geq 27 \cdot (\tan A \cdot \tan B \cdot \tan C)$$
$$\therefore \quad \tan A \cdot \tan B \cdot \tan C \geq 3\sqrt{3}$$

QED

While most triangular trigonometric inequalities are proved by applying AM-GM on their corresponding identities, some may involve other techniques. For instance, the next example utilizes the Euler's formula.

Example 6.5.2

Given any $\triangle ABC$, show that

$$\cos A + \cos B + \cos C \leq \frac{3}{2}$$

Proof

Its corresponding identities are

$$\cos A + \cos B + \cos C = 1 + 4\sin\frac{A}{2}\sin\frac{B}{2}\sin\frac{C}{2}$$

and

$$\sin\frac{A}{2}\sin\frac{B}{2}\sin\frac{C}{2} = \frac{r}{4R}$$

This first identity appears in this chapter's practice. Here, let's assume it holds. The 2nd relation is *(6.8)* on *page 64*. Hence, we have
$$\cos A + \cos B + \cos C = 1 + \frac{r}{R} \qquad (6.14)$$
Euler's formula states that[1]
$$|OI|^2 = R^2 - 2Rr$$
Therefore,
$$R^2 - 2Rr \geq 0 \implies R \geq 2r \implies \frac{r}{R} \leq \frac{1}{2}$$
Setting this to *(6.14)* leads to the desired result immediately.

QED

The preceding problem can also be solved by applying the law of cosines so that it becomes an algebraic inequality with respect to a, b, and c. However, a trigonometric based approach is usually more straightforward if corresponding identities can be determined.

In addition to AM-GM inequality, Cauchy-Schwzarz inequality is also widely used in proving triangular inequality. It is required to solve some practice problems.

Given complex numbers $u_1, u_2, \cdots, u_n$ and $v_1, v_2, \cdots, v_n$, the Cauchy-Schwarz inequality asserts that
$$|u_1v_1 + \cdots + u_nv_n|^2 \leq (|u_1|^2 + \cdots + |u_n|^2)(|v_1|^2 + \cdots + |v_n|^2) \qquad (6.15)$$

6.6 Practice

---

[1] where O and I are the circumcenter and incenter, respectively.

Chapter 6: Trigonometry in Triangle

Practice 1

In $\triangle ABC$, if $\frac{a}{b} = 2 + \sqrt{3}$ and $\angle C = 60°$, find the measurement of $\angle A$ and $\angle B$.

Practice 2

Given $\triangle ABC$, show that
$$\frac{a}{b+c} \geq \sin \frac{A}{2}$$

Practice 3

Given $\triangle ABC$, show that
$$\cos A + \cos B + \cos C = 1 + 4 \sin \frac{A}{2} \sin \frac{B}{2} \sin \frac{C}{2}$$

Practice 4

In $\triangle ABC$, show that
$$\cot A \cot B + \cot B \cot C + \cot C \cot A = 1$$

Practice 5

Show that
$$\tan \frac{A}{2} \tan \frac{B}{2} + \tan \frac{B}{2} \tan \frac{C}{2} + \tan \frac{C}{2} \tan \frac{A}{2} = 1$$

Practice 6

In $\triangle ABC$, show that

$$\tan\frac{A}{2}\tan\frac{B}{2}\tan\frac{C}{2} \leq \frac{\sqrt{3}}{9}$$

Practice 7

In $\triangle ABC$, show that

$$\sin 2A + \sin 2B + \sin 2C = 4\sin A \sin B \sin C$$
$$\cos 2A + \cos 2B + \cos 2C = -1 - 4\cos A \cos B \cos C$$

Practice 8

In $\triangle ABC$, show that

$$\sin^2 A + \sin^2 B + \sin^2 C = 2 + 2\cos A \cos B \cos C$$
$$\cos^2 A + \cos^2 B + \cos^2 C = 1 - 2\cos A \cos B \cos C$$

Practice 9

Given $\triangle ABC$, show that

$$\cos A \cos B \cos C \leq \frac{1}{8}$$
$$\cos^2 A + \cos^2 B + \cos^2 C \geq \frac{3}{4}$$

Chapter 6: Trigonometry in Triangle

Practice 10

In $\triangle ABC$, show that
$$2R \sin A \sin B \sin C = r(\sin A + \sin B + \sin C)$$
where R is the circumradius and r is the inradius.

Practice 11

In $\triangle ABC$, $\angle C = \angle A + 60°$. If $BC = 1$, $AC = r$ and $AB = r^2$, where $r > 1$, prove $r \leq \sqrt{2}$.

Practice 12

Prove that there is one and only one triangle whose side lengths are consecutive integers, and one of whose angles is twice as large as another.
(IMO 1968)

Chapter 7

Additional Techniques

Trigonometry is a powerful and versatile tool. It can also be used to solve some non-trigonometry problems. This chapter discusses several commonly seen such types of non-trigonometry problems. The key to solve such problems is the ability to identify the intrinsic similarities and links between such problems and their corresponding trigonometry tools.

7.1 Applying Law of Cosines

The law of cosines states

$$c^2 = a^2 + b^2 - 2ab\cos C \implies c = \sqrt{a^2 + b^2 - 2ab\cos C}$$

As such, it is a potential candidate to solve a problem involving radical expressions such as

$$\sqrt{a^2 + b^2 - 2ab\cos C} \iff \sqrt{x - y\cos\theta}$$

These two forms are equivalent because a and b are real numbers, so some terms can be consolidated and simplified. The appropriateness of using law of cosines will become more apparent if the given

Chapter 7: Additional Techniques

problem involves the sum of two such radical expressions, i.e.

$$\sqrt{x + y\cos\theta} + \sqrt{m + n\cos\varphi}$$

In this case, the two terms may be viewed as the two sides of a triangle. Let's revisit a practice problem from *Chapter 2* to demonstrate the application of this technique.

Example 7.1.1

Find an acute angle α so that

$$\sqrt{15 - 12\cos\alpha} + \sqrt{7 - 4\sqrt{3}\sin\alpha} = 4$$

Solution

The given equation can be re-written as

$$\sqrt{(\sqrt{12})^2 + (\sqrt{3})^2 - 2\cdot\sqrt{12}\cdot\sqrt{3}\cdot\cos\alpha} \quad (7.1)$$
$$+ \sqrt{(2)^2 + (\sqrt{3})^2 - 2\cdot 2\cdot\sqrt{3}\cdot\cos(90° - \alpha)} \quad (7.2)$$
$$= 4$$

Construct a $Rt\triangle ABC$ as shown below where $CA = \sqrt{12}$ and $BC = 2$. Locate point E so that $CE = \sqrt{3}$, $\angle ACE = \alpha$ and $\angle ECB = 90° - \alpha$.

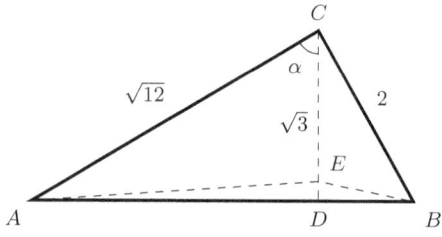

Then, by the Law of Cosines, *(7.1)* equals AE and *(7.2)* equals BE. Hence, we have $AE + BE = 4$. However, by Pythagorean

Chapter 7: Additional Techniques

theorem, $AB = \sqrt{AC^2 + BC^2} = 4$. Therefore, point E must lie on AB. Let it be point D. Now, let's compute α using the area method.

$$S_{\triangle ABC} = S_{\triangle ACD} + S_{\triangle BCD}$$
$$\frac{1}{2} \cdot AC \cdot BC = \frac{1}{2} \cdot AC \cdot DC \cdot \sin\alpha + \frac{1}{2} \cdot DC \cdot BC \sin(90° - \alpha)$$
$$\frac{1}{2} \cdot \sqrt{12} \cdot 2 = \frac{1}{2} \cdot \sqrt{12} \cdot \sqrt{3} \cdot \sin\alpha + \frac{1}{2} \cdot \sqrt{3} \cdot 2 \cdot \cos\alpha$$
$$2 = \sqrt{3}\sin\alpha + \cos\alpha$$
$$2 = 2 \cdot \left(\frac{\sqrt{3}}{2} \cdot \sin\alpha + \frac{1}{2} \cdot \cos\alpha\right)$$
$$2 = 2\sin(30° + \alpha)$$
$$1 = \sin(30° + \alpha)$$
$$\therefore \quad \alpha = \boxed{60°}$$

Done.

The solution also utilizes the $a\sin\theta + b\cos\theta = \sqrt{a^2 + b^2}\sin(\theta + \varphi)$ transformation which is discussed in *Section 4.1*.

Comparing with the conventional solution given in *Chapter 2*, this solution reveals the intrinsic geometry meaning of this problem.

Please note that it is unnecessary to have trigonometric functions explicitly appear inside the radical in order to apply this technique. For example, the following expression

$$\sqrt{x^2 + y^2 - xy} \qquad (7.3)$$

can be rewritten as

$$\sqrt{x^2 + y^2 - 2xy\cos 60°}$$

This means that *(7.3)* represents the 3rd side of a triangle whose two sides are x and y with an included angles of $60°$. Some practice problems in this chapter can be solved using this interpretation.

Chapter 7: Additional Techniques

7.2 Substituting $(a^2 \pm b^2 = r^2)$

When a problem involves an expression in the form of
$$a^2 + b^2 = r^2$$
it may be possible to use the following substitution:
$$a = r\cos\theta \quad \text{and} \quad b = r\sin\theta$$
Similarly, when an expression of $a^2 - b^2 = r^2$ is involved, the following substitution can be employed:
$$a = r\sec\theta \quad \text{and} \quad b = r\tan\theta$$
Let's consider a well-known inequality as an example.

Example 7.2.1

Let a and b be two real numbers. Show that $a^2 + b^2 \geq 2ab$.

Proof

Let $a^2 + b^2 = r^2$, then $a = r\cos\theta$ and $b = r\sin\theta$. It follows that
$$2ab = 2(r\cos\theta)(r\sin\theta) = r^2(2\sin\theta\cos\theta) = r^2\sin 2\theta \leq r^2 = a^2 + b^2$$

QED

7.3 Substitution by $\tan\theta$

The range of the tangent function is entire $\mathbb{R}$. Therefore, given any real number x, it is always possible to find a θ so that $x = \tan\theta$. This means that it is possible to transform an algebraic expression with respect to real numbers to a trigonometric expression which may lead to a cleaner and easier solution.

Let's consider an example.

Chapter 7: Additional Techniques

Example 7.3.1

Let $x_1 = 2017$ and $x_{n+1} = \frac{1+x_n}{1-x_n}$ for every positive integer n. Compute the value of x_{2017}.

Trying a few numbers will reveal that terms of sequence $\{x_n\}$ repeat every four numbers. This intrinsic property can be readily explained by the $\tan \theta$ substitution.

Solution

The given recurring $x_{n+1} = \frac{1+x_n}{1-x_n}$ resembles the tangent sum of angle formula after applying the reverse construction technique (see (4.4) on *page 43*). Therefore, let $x_n = \tan \alpha_n$. It then follows:

$$\tan \alpha_{n+1} = \frac{1 + \tan x_n}{1 - \tan x_n} = \frac{\tan 45° + \tan x_n}{1 - \tan 45° \cdot \tan \alpha_n} = \tan(45° + \alpha_n)$$

Because the tangent function has a period of 180°, we have

$$\begin{aligned}\tan \alpha_n &= \tan(\alpha_{n-1} + 45°) \\ &= \tan(\alpha_{n-2} + 2 \times 45°) \\ &= \tan(\alpha_{n-3} + 3 \times 45°) \\ &= \tan(\alpha_{n-4} + 4 \times 45°) \\ &= \tan \alpha_{n-4}\end{aligned}$$

This proves that the value of x_n repeats every four numbers which means

$$x_{2017} = x_{2013} = \cdots = x_1 = \boxed{2017}$$

Done.

It should be clear by now that trigonometric substitution can be very flexible. In fact, the choice of function is not restricted to $\sin \theta$ or $\tan \theta$. It can be anything which is appropriate to the formation of the given expression. Let's review another example.

Chapter 7: Additional Techniques

Example 7.3.2

Let m be a positive integer. Show that

$$\frac{1}{\sqrt{m+1}} < \sin \frac{1}{\sqrt{m}}$$

Solution

Let $m = \cot^2 \alpha$ where $\alpha \in (0, \frac{\pi}{2})$. Then

$$\frac{1}{\sqrt{m+1}} < \sin \frac{1}{\sqrt{m}} \Leftrightarrow \sin \alpha < \sin \tan \alpha$$

By *(2.11)* on *page 12*, this relation will hold if $0 < \tan \alpha < \frac{\pi}{2}$. This is indeed the case because

$$\cot^2 \alpha = m \geq 1 \implies \tan \alpha \leq 1 < \frac{\pi}{2}$$

Done.

7.4 Converse of Triangular Identities

This is an advanced technique even though essentially it is still the reverse construction method which is discussed in *Chapter 4*. The difference is that the technique discussed here is to reverse a triangular identity instead of simply converting a real number to a trigonometric function.

For example, given the identity *(6.10)* on *page 67*:

$$\tan A + \tan B + \tan C = \tan A \cdot \tan B \tan C$$

we can claim that if three real numbers x, y, z satisfy the relation $x + y + z = xyz$, then there must exist a triangle $\triangle ABC$ such that

Chapter 7: Additional Techniques

$\tan A = x$, $\tan B = y$ and $\tan C = z$. Furthermore, if x, y, z are known to be positive, then $\triangle ABC$ must be acute.

This conclusion can be proved by showing $A + B + C = 180°$ which can use a similar approach as to prove the original trigonometric identity *(6.10)*.

Let's consider an example of applying this technique.

Example 7.4.1

Let x, y, and z be three real numbers such that $x + y + z = xyz$. Show
$$\frac{x}{\sqrt{1+x^2}} + \frac{y}{1+y^2} + \frac{1}{\sqrt{1+z^2}} \leq \frac{3\sqrt{3}}{2}$$

Proof

Because $x + y + z = xzy$, there must exist a triangle $\triangle ABC$ such that $\tan A = x$, $\tan B = y$ and $\tan C = z$. Then,

$$\frac{x}{\sqrt{1+x^2}} = \frac{\tan A}{\sqrt{1+\tan^2 A}} = \frac{\tan A}{\sec A} = \sin A$$

Similarly, the other two terms are equivalent to $\sin B$ and $\sin C$, respectively. Hence, the to be proved claim is equivalent to

$$\sin A + \sin B + \sin C \leq \frac{3\sqrt{3}}{2}$$

This inequality is proved in the previous chapter's practice.

QED

Chapter 7: Additional Techniques

7.5 Practice

Practice 1

Let real numbers x and y satisfy the relation $4x^2 - 5xy + 4y^2 = 5$. Find the maximum and minimal value of $x^2 + y^2$.

Practice 2

Given non-negative real numbers x, y and z, prove

$$\sqrt{x^2 + y^2 - xy} + \sqrt{y^2 + z^2 - yz} \geq \sqrt{x^2 + z^2 + xz}$$

Practice 3

Solve this inequality

$$\frac{x}{\sqrt{x^2 + 1}} + \frac{1 - x^2}{1 + x^2} > 0$$

Practice 4

Given any five real numbers, show that at least two of them x and y satisfy the condition $|xy + 1| > |x - y|$.

Chapter 7: Additional Techniques

Practice 5

Let $\{x_n\}$ and $\{y_n\}$ be two real number sequences which are defined as follow:

$$x_1 = y_1 = \sqrt{3}, \quad x_{n+1} = x_n + \sqrt{1 + x_n^2}, \quad y_{n+1} = \frac{y_n}{1 + \sqrt{1 + y_n^2}}$$

for all $n \geq 1$. Prove that $2 < x_n y_n < 3$ for all $n > 1$.

Practice 6

Let x, y, z be three positive real numbers satisfying $xyz + x + z = y$. Find the maximum value of

$$P = \frac{2}{x^2 + 1} - \frac{2}{y^2 + 1} + \frac{3}{z^2 + 1}$$

Practice 7

Let m be a positive integer. Show that

$$\sin \frac{2}{\sqrt{m}} < \frac{2}{\sqrt{m+1}}$$

Practice 8

Prove

$$\frac{1}{\sqrt{2019}} < \underbrace{\sin \sin \sin \cdots \sin}_{2017} \frac{\sqrt{2}}{2} < \frac{2}{\sqrt{2019}}$$

83

Chapter 7: Additional Techniques

Practice 9

Solve this equation

$$2\sqrt{2}x^2 + x - \sqrt{1-x^2} - \sqrt{2} = 0$$

Practice 10

Let a and b be two positive real numbers not exceeding 1. Prove

$$\frac{1}{\sqrt{a^2+1}} + \frac{1}{\sqrt{b^2+1}} \leq \frac{2}{\sqrt{1+ab}}$$

(Russia)

Chapter 8

Solutions

8.1 Introduction

This section is intentionally left blank.

So section numbers of solutions and practices can match.

Chapter 8: Solutions

8.2 Trigonometry Basics

Practice 1

Compute the following values:

i) $\cos 75°$

ii) $\sin 165°$

iii) $\sin 105°$

i) $\cos 75° = \cos(90° - 15°) = \sin 15° = \frac{\sqrt{6}-\sqrt{2}}{4}$.

ii) $\sin 165° = \sin(180° - 15°) = \sin 15° = \frac{\sqrt{6}-\sqrt{2}}{4}$.

iii) $\sin 105° = \sin(90° + 15°) = \cos 15° = \frac{\sqrt{6}+\sqrt{2}}{4}$.

Practice 2

Show that
$$\sec^2 \alpha = 1 + \tan^2 \alpha$$
$$\csc^2 \alpha = 1 + \cot^2 \alpha$$

Typically, such simple relations can be derived from basic definitions. Let's prove the first one starting from right side and the second one from the left side.

$$1 + \tan^2 \alpha = 1 + \frac{\sin^2 \alpha}{\cos^2 \alpha} = \frac{\cos^2 \alpha + \sin^2 \alpha}{\cos^2 \alpha} = \frac{1}{\cos^2 \alpha} = \sec^2 \alpha$$

and

$$\csc^2 \alpha = \frac{1}{\sin^2 \alpha} = \frac{\sin^2 \alpha + \cos^2 \alpha}{\sin^2 \alpha} = 1 + \frac{\cos^2 \alpha}{\sin^2 \alpha} = 1 + \cot^2 \alpha$$

Note: when proving the 2nd relation, we transformed a number,

Chapter 8: Solutions

i.e. 1, to an trigonometric expression, i.e. $(\sin^2 \alpha + \cos^2 \alpha)$. This turns out to be a useful technique. This technique will be discussed in more detail in *Chapter 4*.

Practice 3

Prove the identity: $\tan^2 x - \sin^2 x = \tan^2 x \sin^2 x$.

This identity can be proved by applying the relevant definition:

$$\begin{aligned}
\tan^2 x - \sin^2 x &= \frac{\sin^2 x}{\cos^2 x} - \sin^2 x \\
&= \left(\frac{1}{\cos^2 x} - 1\right) \sin^2 x \\
&= (\sec^2 x - 1) \sin^2 x \\
&= \tan^2 x \sin^2 x \qquad \because (2.17) \text{ on page 14}
\end{aligned}$$

Practice 4

Compute $\sin 15°$ and $\cos 15°$ using a geometry approach.

Construct a right $\triangle ABC$ where $\angle A = 30°$ and $\angle C = 90°$. Let the angle bisector of A meet BC at D, then $\angle DAC = 15°$.

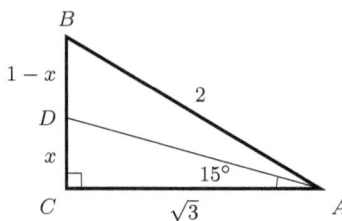

Let $AB = 2$, then by the property of $30° - 60° - 90°$ right triangle, BC must equal 1 and $AC = \sqrt{3}$. Furthermore, assuming

Chapter 8: Solutions

$CD = x$, then $BD = 1 - x$. By the angle bisector theorem, we have

$$\frac{BD}{CD} = \frac{AB}{AC} \implies \frac{1-x}{x} = \frac{2}{\sqrt{3}} \implies x = \frac{\sqrt{3}}{2+\sqrt{3}} = 2\sqrt{3} - 3$$

Applying the Pythagorean theorem on $\triangle ACD$ leads

$$AD^2 = AC^2 + CD^2 = 24 - 12\sqrt{3} \implies AD = \sqrt{6} \cdot (\sqrt{3} - 1)$$

Now, we have

$$\sin 15° = \frac{CD}{AD} = \frac{2\sqrt{3} - 3}{\sqrt{6} \cdot (\sqrt{3} - 1)} = \frac{\sqrt{6} - \sqrt{2}}{4}$$

$$\cos 15° = \frac{AC}{AD} = \frac{\sqrt{3}}{\sqrt{6} \cdot (\sqrt{3} - 1)} = \frac{\sqrt{6} + \sqrt{2}}{4}$$

Practice 5

Let a and b be the two sides of a triangle, and their included angle be C. Show that the area of this triangle equals

$$S = \frac{1}{2} \cdot ab \sin C \tag{8.1}$$

As shown in the diagram below, the altitude from vertex A equals $(b \sin C)$. Therefore, by the base-altitude formula, we have

$$S = \frac{1}{2} \cdot a \cdot (b \sin C) = \frac{1}{2} \cdot ab \sin C$$

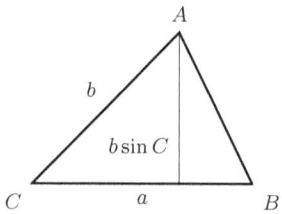

Practice 6

If $\cos x - \sin x = \sqrt{2}\sin x$, prove $\cos x + \sin x = \sqrt{2}\cos x$.

From the given condition, we have

$$\cos x = (\sqrt{2}+1)\sin x \implies \sin x = \frac{1}{\sqrt{2}+1}\cos x = (\sqrt{2}-1)\cos x$$

$$\therefore \quad \cos x + \sin x = \cos x + (\sqrt{2}-1)\cos x = \sqrt{2}\cos x$$

Practice 7

Find an acute angle α so that

$$\sqrt{15-12\cos\alpha} + \sqrt{7-4\sqrt{3}\sin\alpha} = 4$$

Let $\sqrt{15-12\cos\alpha} = a$ and $\sqrt{7-4\sqrt{3}\sin\alpha} = b$, then $a+b=4$.

It follows that
$$\cos\alpha = \frac{15-a^2}{12}$$

and
$$\sin\alpha = \frac{7-b^2}{4\sqrt{3}} = \frac{7-(4-a)^2}{4\sqrt{3}}$$

Because $\cos^2\alpha + \sin^2\alpha = 1$, therefore it must hold

$$\left(\frac{15-a^2}{12}\right)^2 + \left(\frac{7-(4-a)^2}{4\sqrt{3}}\right)^2 = 1$$

$$\Leftrightarrow a^4 - 12a^3 + 54a^2 - 108a + 81 = 0$$
$$\Leftrightarrow (a-3)^4 = 0$$
$$\Leftrightarrow a = 3$$

Chapter 8: Solutions

$$\therefore \quad \sqrt{15 - 12\cos\alpha} = 3 \implies \cos\alpha = \frac{1}{2} \implies \alpha = \boxed{60°}$$

Note: this problem can be solved by an alternative approach which is presented in *Section 7.1*.

Practice 8

Find the range of real number a if the following equation of x is solvable in real numbers:

$$\sin^2 x + \cos x + a = 0$$

The given relation can be rewritten as a quadratic equation with respect to $\cos x$ by replacing $\sin^2 x$ with $(1 - \cos^2 x)$:

$$\cos^2 x - \cos x - (a+1) = 0 \tag{8.2}$$

For *(8.2)* to be solvable, its determinant must be non-negative:

$$(-1)^2 + 4 \cdot (a+1) \geq 0 \implies a \geq -\frac{5}{4}$$

Meanwhile, because

$$-1 \leq \cos x = \frac{1 \pm \sqrt{1 + 4 \cdot (a+1)}}{2} \leq 1 \implies -\frac{5}{4} \leq a \leq 1$$

Therefore, the answer is $\boxed{-\frac{5}{4} \leq a \leq 1}$.

Alternative Solution

Equation 8.2 can be rewritten as

$$a = \cos^2 x - \cos x - 1 = \left(\cos x - \frac{1}{2}\right)^2 - \frac{5}{4}$$

Therefore, the minimal value of a is $0 - \frac{5}{4} = -\frac{5}{4}$ and the maximum value equals $\left(-1 - \frac{1}{2}\right)^2 - \frac{5}{4} = 1$.

Chapter 8: Solutions

Practice 9

Find the number of solutions to the equation $\sin x = \frac{x}{2018}$.

This is equivalent to counting the intersection points of line $y = \frac{x}{2018}$ and the plot $y = \sin x$.

As shown in the graph below, including 0, there are two intersection points in $[0, \pi]$, two in $[2\pi, 3\pi]$, etc, on the positive x side. This means there are two intersection points in every 2π interval. When the straight line $y = \frac{x}{2018}$ goes beyond $y = 1$ (correspondingly, $x = 2018$), there will be no more intersection point.

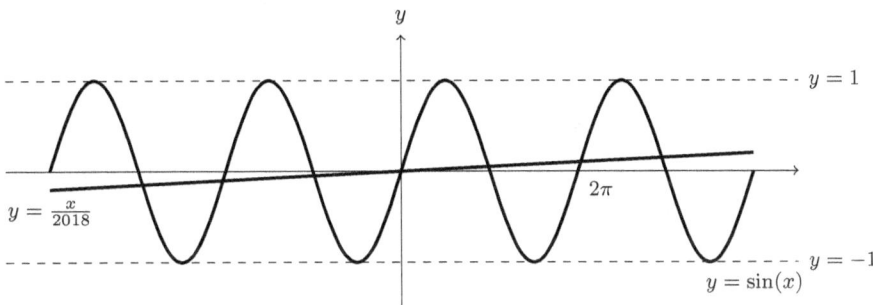

Because $2018 \div (2\pi) \approx 321.17$, there will be 321 whole 2π waves. This means there will be $321 \times 2 = 642$ points. The additional 0.17 wave will not intersect the line $y = \frac{x}{2018}$ because the line will be very close to the line $y = 1$ but 0.17 wave will not reach that high.

The number of intersection points on the negative side is exactly the same as that on the positive side. However, the point 0 is double counted. Therefore, the final answer is $642 \times 2 - 1 = \boxed{1283}$.

Chapter 8: Solutions

Practice 10

Prove that function $f(x) = \cos \sqrt{x}$ is not a periodical function.

If this claim is not true, then there exists a constant positive real number T such that for every real number x, (8.3) below always hold.
$$\cos \sqrt{x+T} = \cos \sqrt{x} \qquad (8.3)$$
Setting $x = 0$ leads to
$$\cos \sqrt{T} = \cos 0 = 1 \implies \sqrt{T} = 2k\pi \qquad (k \in \mathbb{Z})$$
Next, setting $x = T$ and also noting $\cos\sqrt{T} = 1$ (see above) yield
$$\cos\sqrt{2T} = \cos\sqrt{T} = 1 \implies \sqrt{2T} = 2m\pi \qquad (m \in \mathbb{Z})$$
$$\therefore \quad \sqrt{T} = \frac{1}{\sqrt{2}} \cdot (2m\pi) = 2k\pi \implies \sqrt{2} = \frac{m}{k}$$

However, this is impossible because $\sqrt{2}$ is irrational, but the right side is rational.

Practice 11

Sort $\sin(-1)$, $\cos(-1)$, and $\tan(-1)$ in an ascending order.

Because $-\frac{\pi}{2} < -1 < 0$, the corresponding angle is in the *IV* quadrant. Hence, $\sin(-1) < 0$, $\cos(-1) > 0$, and $\tan(-1) < 0$. Meanwhile,
$$|\cos(-1)| < 1 \implies |\tan(-1)| = \left|\frac{\sin(-1)}{\cos(-1)}\right| > |\sin(-1)|$$
$$\therefore \quad \tan(-1) < \sin(-1) < \cos(-1)$$

Practice 12

Let $0° < \alpha < 45°$, explain that

$$(\tan\alpha)^{\cot\alpha} < (\tan\alpha)^{\tan\alpha} < (\cot\alpha)^{\tan\alpha} < (\cot\alpha)^{\cot\alpha}$$

This problem can be solved by noting that tangent function is increasing in $(0, 45°)$ and the cotangent function is decreasing in the same range.

In this range, we have $\tan\alpha < 1 < \cot\alpha$. When $a < 1$, function a^x is decreasing, therefore

$$(\tan\alpha)^{\cot\alpha} < (\tan\alpha)^{\tan\alpha}$$

Similarly, when $a > 1$, function a^x is increasing. Hence,

$$(\cot\alpha)^{\tan\alpha} < (\cot\alpha)^{\cot\alpha}$$

Now, when $0 < x < 1$, function x^a is increasing. Therefore,

$$(\tan\alpha)^{\tan\alpha} < (\cot\alpha)^{\tan\alpha}$$

Combining these three relations yields the desired claim.

Practice 13

Let $\theta \in [0, 2\pi]$ satisfying

$$\cos^5\theta - \sin^5\theta < 7(\sin^3\theta - \cos^3\theta)$$

Find the range of θ.

The given relation is equivalent to

$$\cos^5\theta + 7\cos^3\theta < \sin^5\theta + 7\sin^3\theta$$

Because the function

$$f(x) = x^5 + 7x^3$$

Chapter 8: Solutions

monotonically increases, therefore we must have

$$\cos\theta < \sin\theta \implies \theta \in \boxed{\left(\frac{\pi}{4}, \frac{5\pi}{4}\right)}$$

Practice 14

If $0 < \alpha < \beta < \frac{\pi}{2}$, show

$$\frac{\cot\beta}{\cot\alpha} < \frac{\cos\beta}{\cos\alpha} < \frac{\beta}{\alpha}$$

The first part of the inequality can be rewritten as

$$\frac{\cot\beta}{\cot\alpha} < \frac{\cos\beta}{\cos\alpha} \Leftrightarrow \frac{\cot\beta}{\cos\beta} < \frac{\cot\alpha}{\cos\alpha} \Leftrightarrow \sin\alpha < \sin\beta$$

This obviously holds because sine function increases in the region of $[0, \frac{\pi}{2}]$ and $0 < \alpha < \beta < \frac{\pi}{2}$.

The second part of the inequality can be rewritten as

$$\frac{\cos\beta}{\cos\alpha} < \frac{\beta}{\alpha} \Leftrightarrow \frac{\cos\beta}{\beta} < \frac{\cos\alpha}{\alpha}$$

This form is essentially to compare the slopes of two lines. To see this, let's take two points on the function $y = \cos x$: $A(\alpha, \cos\alpha)$ and $B(\beta, \cos\beta)$.

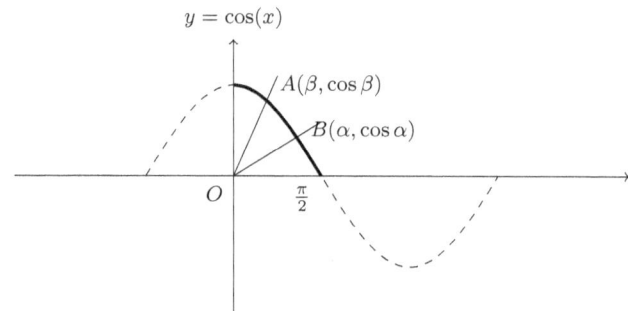

Because cosine function decreases in $[0, \frac{\pi}{2}]$ and $0 < \alpha < \beta < \frac{\pi}{2}$, therefore point A is above point B. Consequently, the slope of line OA is larger than that of OAB:

$$k_{OA} > k_{OB} \implies \frac{\cos \alpha}{\alpha} > \frac{\cos \beta}{\beta}$$

Chapter 8: Solutions

8.3 Trigonometric Identities

Practice 1

Show that $\sin\alpha - \sin\beta = 2\sin\frac{\alpha-\beta}{2}\cos\frac{\alpha+\beta}{2}$.

This identity can be proved in a similar way as *Example 3.4.1* on *page 28*. Alternatively, it can be proved using *(3.22)* on *page 27* together with the fact that sine function is an odd function.

$$\sin\alpha - \sin\beta = \sin\alpha + \sin(-\beta)$$
$$= 2\sin\frac{\alpha+(-\beta)}{2}\cos\frac{\alpha-(-\beta)}{2}$$
$$= 2\sin\frac{\alpha-\beta}{2}\cos\frac{\alpha+\beta}{2}$$

Practice 2

If $\sin x = \frac{2}{5}$ and x is acute, compute the values of

i) $\cos 2x$.

ii) $\cos 4x$.

iii) $\sin 2x$.

iv) $\sin 4x$.

i) $\cos 2x = 1 - 2\sin^2 x = 1 - 2\cdot(\frac{2}{5})^2 = \frac{17}{25}$.

ii) $\cos 4x = 2\cos^2 2x - 1 = 2\cdot(\frac{17}{25})^2 - 1 = -\frac{47}{625}$.

iii) $\sin 2x = 2\sin x \cos x = 2\cdot\frac{2}{5}\cdot\sqrt{1-(\frac{2}{5})^2} = \frac{4}{25}\sqrt{21}$.

iv) $\sin 4x = \sqrt{1-\cos^2 4x} = \sqrt{1-(-\frac{47}{625})^2} = \frac{136}{625}\sqrt{21}$.

The reason that $\sin 4x$ is positive is because $\sin x = \frac{2}{5} < \frac{\sqrt{2}}{2}$ means $x < 45°$ or $4x < 180°$.

Practice 3

Let $\alpha \in \left(\frac{3\pi}{2}, 2\pi\right)$. Simplify

$$\sqrt{\frac{1}{2} - \frac{1}{2}\sqrt{\frac{1}{2} + \frac{1}{2} \cdot \cos 2\alpha}}$$

By the half-angle formula (3.9) and (3.10) on page 23.

$$\alpha \in \left(\frac{3\pi}{2}, 2\pi\right) \implies \sqrt{\frac{1}{2} + \frac{1}{2} \cdot \cos 2\alpha} = |\cos \alpha| = \cos \alpha$$

also

$$\frac{\alpha}{2} \in \left(\frac{3\pi}{4}, \pi\right) \implies \sqrt{\frac{1}{2} - \frac{1}{2} \cdot \cos \alpha} = \left|\sin \frac{\alpha}{2}\right| = \sin \frac{\alpha}{2}$$

Hence the final result is $\boxed{\sin \frac{\alpha}{2}}$.

Practice 4

Prove the following identity

$$\tan \alpha + \tan(90° - \alpha) = \frac{2}{\sin 2\alpha} \qquad (8.4)$$

By the definition, we have $\tan(90° - \alpha) = \cot \alpha$. Therefore,

$$\tan \alpha + \tan(90° - \alpha)$$
$$= \tan \alpha + \cot \alpha$$
$$= \frac{\sin \alpha}{\cos \alpha} + \frac{\cos \alpha}{\sin \alpha}$$

Chapter 8: Solutions

$$= \frac{\sin^2\alpha + \cos^2\alpha}{\sin\alpha\cos\alpha}$$

$$= \frac{1}{\frac{1}{2}\sin 2\alpha}$$

$$= \frac{2}{\sin 2\alpha}$$

Practice 5

Prove
$$\frac{\sin(\alpha+\beta)}{\sin(\alpha-\beta)} = \frac{1+\cot\alpha\tan\beta}{1-\cot\alpha\tan\beta}$$

Applying the sum and difference formulas:

$$\frac{\sin(\alpha+\beta)}{\sin(\alpha-\beta)} = \frac{\sin\alpha\cos\beta + \cos\alpha\sin\beta}{\sin\alpha\cos\beta - \cos\alpha\sin\beta}$$

Then dividing both the numerator and the denominator by $\sin\alpha\cos\beta$ yields the desired result.

Practice 6

Prove
$$\frac{\cos(\alpha+\beta)}{\cos(\alpha-\beta)} = \frac{1-\tan\alpha\tan\beta}{1+\tan\alpha\tan\beta}$$

Applying the sum and difference formulas:

$$\frac{\cos(\alpha+\beta)}{\cos(\alpha-\beta)} = \frac{\cos\alpha\cos\beta - \sin\alpha\sin\beta}{\cos\alpha\cos\beta + \sin\alpha\sin\beta}$$

Then dividing both the numerator and the denominator by $\cos\alpha\cos\beta$ yields the desired result.

Practice 7

Solve the following equation for $0 \leq x \leq 2\pi$:

$$\sin x + \sin \frac{x}{2} = 0$$

The given equation is equivalent to:

$$2 \sin \frac{x}{2} \cos \frac{x}{2} + \sin \frac{x}{2} = 0$$
$$\sin \frac{x}{2} \left(2 \cos \frac{x}{2} + 1\right) = 0$$

$$\therefore \quad \sin \frac{x}{2} = 0 \quad \text{or} \quad \cos \frac{x}{2} = -\frac{1}{2}$$

When $0 \leq x \leq 2\pi$, then $0 \leq \frac{x}{2} \leq \pi$.

$\sin \frac{x}{2} = 0 \implies \frac{x}{2} = 0, \pi \implies x = 0, 2\pi$.

$\cos \frac{x}{2} = -\frac{1}{2} \implies \frac{x}{2} = \frac{2\pi}{3} \implies x = \frac{4\pi}{3}$.

Therefore, all the solutions are $\boxed{0, \frac{4\pi}{3}, 2\pi}$.

Practice 8

Prove the following identity:

$$\sin \alpha + \sin \beta + \sin \gamma - \sin(\alpha + \beta + \gamma)$$
$$= 4 \sin \frac{\alpha + \beta}{2} \sin \frac{\beta + \gamma}{2} \sin \frac{\gamma + \alpha}{2} \qquad (8.5)$$

Because the left side is a sum of several terms and the right side is a product, let's natural to apply the sum-product transformation.

$$(\sin \alpha + \sin \beta) + (\sin \gamma - \sin(\alpha + \beta + \gamma))$$

Chapter 8: Solutions

$$=2\sin\frac{\alpha+\beta}{2}\cos\frac{\alpha-\beta}{2}+2\sin\frac{\gamma-(\alpha+\beta+\gamma)}{2}\cos\frac{\gamma+(\alpha+\beta+\gamma)}{2}$$

$$=2\sin\frac{\alpha+\beta}{2}\left(\cos\frac{\alpha-\beta}{2}-\cos\frac{2\gamma+\alpha+\beta}{2}\right)$$

$$=2\sin\frac{\alpha+\beta}{2}\cdot\left(-2\sin\frac{(\alpha-\beta)+(2\gamma+\alpha+\beta)}{4}\sin\frac{(\alpha-\beta)-(2\gamma+\alpha+\beta)}{4}\right)$$

$$=4\sin\frac{\alpha+\beta}{2}\sin\frac{\beta+\gamma}{2}\sin\frac{\gamma+\alpha}{2}$$

Quiz: what if $\alpha+\beta+\gamma=\pi$?

Practice 9

Show that if $m\tan(\theta-30°)=n\tan(\theta+120°)$, then

$$\cos 2\theta=\frac{m+n}{2(m-n)}$$

Expanding the given relation yields:

$$m\cdot\frac{\tan\theta-\tan 30°}{1+\tan\theta\tan 30°}=n\cdot\frac{\tan\theta+\tan 120°}{1-\tan\theta\tan 120°}$$

Let $t=\tan\theta$, and also evaluate $\tan 30°$ and $\tan 120°$:

$$m\cdot\frac{t-\frac{1}{\sqrt{3}}}{1+\frac{t}{\sqrt{3}}}=n\cdot\frac{t+\sqrt{3}}{1-t\cdot\sqrt{3}}$$

Solving this relation leads to

$$t^2=\frac{m-3n}{3m-n}$$

Now by (4.7) on page 46:

$$\cos 2\theta=\frac{1-t^2}{1+t^2}=\frac{1-\frac{m-3n}{3m-n}}{1+\frac{m-3n}{3m-n}}=\frac{m+n}{2(m-n)}$$

Practice 10

Compute the value of $(\tan 9° - \tan 27° - \tan 63° + \tan 81°)$.

By *(8.4)* on *page 97*, we have

$$\tan 9° + \tan 81° = \frac{2}{\sin 18°} = \frac{2 \times 4}{\sqrt{5} - 1} = 2 \times (\sqrt{5} + 1)$$

and

$$\tan 27° + \tan 63° = \frac{2}{\sin 54°} = \frac{2}{\cos 36°} = \frac{2}{1 - 2\sin^2 18°}$$
$$= \frac{2}{1 - 2 \times (\frac{\sqrt{5}-1}{4})^2} = 2 \times (\sqrt{5} - 1)$$

Therefore, the original expression equals

$$2 \times (\sqrt{5} + 1) - 2 \times (\sqrt{5} - 1) = \boxed{4}$$

Practice 11

Compute $\sin 25° \sin 35° \sin 85°$.

This is a typical problem that can be solved by applying sum-to-product and product-to-sum formulas:

$$\sin 25° \sin 35° \sin 85°$$
$$= \frac{1}{2} \cdot (\cos 10° - \cos 60°) \cdot \sin 85°$$
$$= \frac{1}{2} \cdot (\cos 10° \sin 85° - \cos 60° \sin 85°)$$
$$= \frac{1}{2} \cdot \cos 10° \sin 85° - \frac{1}{2} \cdot \cos 60° \sin 85°$$
$$= \frac{1}{4} \cdot (\sin 95° + \sin 75°) - \frac{1}{4} \cdot \sin 85°$$
$$= \frac{1}{4} \cdot (\sin 85° + \cos 15°) - \frac{1}{4} \cdot \sin 85°$$

Chapter 8: Solutions

$$= \frac{1}{4} \cdot \cos 15°$$
$$= \frac{\sqrt{6} + \sqrt{2}}{16}$$

While applying the sum-product formula is the "standard" approach, this problem can also be solved by applying the triple angle formula *(3.15)* on *page 26* by setting α to 25°.

Practice 12

Prove $\tan 20° \tan 40° \tan 60° \tan 80° = 3$.

Because $\tan 60° = \sqrt{3}$, the given problem is equivalent to show

$$\tan 20° \tan 40° \tan 80° = \sqrt{3} \tag{8.6}$$

This relation indeed holds by the triple tangent formula *(3.17)* on *page 26* where $\alpha = 20°$.

Practice 13

Show that $\cot 70° + 4\cos 70° = \sqrt{3}$.

This is a typical problem that can be solved by employing various identities.

$$\cot 70° + 4\cos 70°$$
$$= \tan 20° + 4\sin 20°$$
$$= \frac{\sin 20°}{\cos 20°} + 4\sin 20°$$
$$= \frac{1}{\cos 20°} \cdot (\sin 20° + 4\sin 20° \cos 20°)$$
$$= \frac{1}{\cos 20°} \cdot (\sin 20° + 2\sin 40°)$$
$$= \frac{1}{\cos 20°} \cdot ((\sin 20° + \sin 40°) + \sin 40°)$$

$$= \frac{1}{\cos 20°} \cdot (2\sin 30° \cos 10° + \sin 40°)$$
$$= \frac{1}{\cos 20°} \cdot (\cos 10° + \sin 40°)$$
$$= \frac{1}{\cos 20°} \cdot (\sin 80° + \sin 40°)$$
$$= \frac{1}{\cos 20°} \cdot 2 \cdot \sin 60° \cos 20°$$
$$= 2\sin 60°$$
$$= \sqrt{3}$$

Practice 14

If $\frac{1+\tan\alpha}{1-\tan\alpha} = 2018$, show that $\sec 2\alpha + \tan 2\alpha = 2018$.

The target is a function of 2α and the given is a function of $\tan\alpha$. Therefore, let's apply the half tangent substitution formulas *(4.7)* and *(4.8)* on *page 46* to simplify the target.

$$\sec 2\alpha + \tan 2\alpha$$
$$= \frac{1}{\cos 2\alpha} + \tan 2\alpha$$
$$= \frac{1+\tan^2\alpha}{1-\tan^2\alpha} + \frac{2\tan\alpha}{1-\tan^2\alpha} \qquad \because (4.8)$$
$$= \frac{(1+\tan\alpha)^2}{1-\tan^2\alpha}$$
$$= \frac{(1+\tan\alpha)^2}{(1+\tan\alpha)(1-\tan\alpha)}$$
$$= \frac{1+\tan\alpha}{1-\tan\alpha}$$
$$= 2018$$

Chapter 8: Solutions

Practice 15

Compute the value of $\cos 6° \cos 42° \cos 66° \cos 78°$.

This expression can be evaluated by applying *(3.16)* twice:

$$\begin{aligned}
&\cos 6° \cos 42° \cos 66° \cos 78° \\
&= \cos 6° \cos 42° \cos 66° \cos 78° \cdot \frac{\cos 54°}{\cos 54°} \\
&= (\cos 6° \cos 54° \cos 66°) \cdot \frac{\cos 42° \cos 78°}{\cos 54°} \\
&= (\cos 6° \cos(60° - 6°) \cos(60° + 6°)) \cdot \frac{\cos 42° \cos 78°}{\cos 54°} \\
&= \frac{\cos(3 \cdot 6°)}{4} \cdot \frac{\cos 42° \cos 78°}{\cos 54°} \\
&= \frac{1}{4} \cdot \frac{1}{\cos 54°} \cdot \cos 18° \cos 42° \cos 78° \\
&= \frac{1}{4} \cdot \frac{1}{\cos 54°} \cdot \frac{1}{4} \cdot \cos(3 \cdot 18°) \\
&= \boxed{\frac{1}{16}}
\end{aligned}$$

Practice 16

In $\triangle ABC$, find the measurement of C if

$$3 \sin A + 4 \cos B = 6 \quad \text{and} \quad 4 \sin B + 3 \cos A = 1$$

We note that both sine and cosine of A and B appear in the given relation and also the coefficients at symmetric. Hence, let's square both equations:

$$9 \sin^2 A + 24 \sin A \cos B + 16 \cos^2 B = 36$$
$$16 \sin^2 B + 24 \sin B \cos A + 9 \cos^2 B = 1$$

Chapter 8: Solutions

Adding these two equations and rearranging leads to:

$$9(\sin^2 A+\cos^A)+16(\cos^2 B+\sin^2 B)+24(\sin A\cos B+\cos A\sin B)=37$$

$$\implies 9+16+24\sin(A+B)=37$$

$$\therefore\quad 24\sin(A+B)=12 \implies \sin(A+B)=\frac{1}{2}$$

This means that $A+B=30°$ or $A+B=150°$. However, if $A+B=30°$, then $A<30°$ which will imply

$$3\sin A+4\cos B<3\sin 30°+4\cos B\leq 3\times\frac{1}{2}+4<6$$

This contradicts to the given condition. Therefore $A+B=150°$ which means $C=\boxed{30°}$.

Practice 17

Compute the value of $(\sin^4 10°+\sin^4 50°+\sin^4 70°)$.

(Tsinghua)

Repeatedly applying *(3.9)* on *page 23*, *(3.10)*, and sum-to-product formula yields:

$$\begin{aligned}
&\sin^4 10°+\sin^4 50°+\sin^4 70° \\
=&\left(\frac{1-\cos 20°}{2}\right)^2+\left(\frac{1-\cos 100°}{2}\right)^2+\left(\frac{1-\cos 140°}{2}\right)^2 \\
=&\frac{3}{4}-\frac{1}{2}\cdot(\cos 20°+\cos 100°+\cos 140°) \\
&+\frac{1}{4}\cdot(\cos^2 20°+\cos^2 100°+\cos^2 140°) \\
=&\frac{3}{4}-\frac{1}{2}\cdot(2\cos 60°\cos 40°-\cos 40°) \\
&+\frac{1}{4}\cdot\left(\frac{1+\cos 40°}{2}+\frac{1+\cos 200°}{2}+\frac{1+\cos 280°}{2}\right) \\
=&\frac{3}{4}-0+\frac{3}{8}+\frac{1}{8}\cdot(\cos 40°-\cos 20°+\cos 80°)
\end{aligned}$$

Chapter 8: Solutions

$$= \frac{9}{8} + \frac{1}{8} \cdot (-2\sin 30° \sin 10° + \sin 10°)$$

$$= \boxed{\frac{9}{8}}$$

Practice 18

Compute the value of $(\sin 1° \sin 2° \cdots \sin 89°)$.

This problem can be solved by repeatedly applying the triple angle formulas *(3.15)* and *(3.16)* on *page 26*:

$$\sin 1° \sin 2° \cdots \sin 89°$$
$$= (\sin 1° \sin 59° \sin 61°)(\sin 2° \sin 58° \sin 62°) \cdots$$
$$(\sin 29° \sin 31°89°) \sin 30° \sin 60°$$
$$= \left(\frac{1}{4}\right)^{29} \sin 3° \sin 6° \cdots \sin 87° \cdot \frac{1}{2} \cdot \frac{\sqrt{3}}{2}$$
$$= \left(\frac{1}{4}\right)^{30} \cdot \sqrt{3} \cdot (\sin 3° \sin 57° \sin 63°)(\sin 6° \sin 54° \sin 66°) \cdots$$
$$(\sin 27° \sin 33° \sin 87°) \sin 30° \sin 60°$$
$$= \left(\frac{1}{4}\right)^{40} \cdot 3 \cdot \sin 9° \sin 18° \cdots \sin 81°$$
$$= \left(\frac{1}{4}\right)^{40} \cdot 3 \cdot (\sin 9° \sin 81°)(\sin 18° \sin 72°)(\sin 27° \sin 63°)$$
$$(\sin 36° \sin 54°) \sin 45°$$
$$= \left(\frac{1}{4}\right)^{40} \cdot 3 \cdot \frac{\sqrt{2}}{2} \cdot (\sin 9° \cos 9°)(\sin 18° \cos 18°)(\sin 27° \cos 27°)$$
$$(\sin 36° \cos 36°)$$
$$= \left(\frac{1}{4}\right)^{40} \cdot 3 \cdot \frac{\sqrt{2}}{2} \cdot \left(\frac{1}{2}\sin 18°\right)\left(\frac{1}{2}\sin 36°\right)\left(\frac{1}{2}\sin 54°\right)\left(\frac{1}{2}\sin 72°\right)$$
$$= \left(\frac{1}{2}\right)^{85} \cdot 3 \cdot \sqrt{2} \cdot \sin 18° \sin 36° \sin 54° \sin 72°$$
$$= \left(\frac{1}{2}\right)^{85} \cdot 3 \cdot \sqrt{2} \cdot \sin 18° \sin 36° \cos 36° \cos 18°$$

$$
\begin{aligned}
&= \left(\frac{1}{2}\right)^{85} \cdot 3 \cdot \sqrt{2} \cdot (\sin 18° \cos 18°)(\sin 36° \cos 36°) \\
&= \left(\frac{1}{2}\right)^{86} \cdot 3 \cdot \sqrt{2}(\sin 18° \cos 18°) \sin 72° \\
&= \left(\frac{1}{2}\right)^{86} \cdot 3 \cdot \sqrt{2}(\sin 18° \cos 18°) \cos 18° \\
&= \left(\frac{1}{2}\right)^{86} \cdot 3 \cdot \sqrt{2} \sin 18° \cos^2 18° \\
&= \left(\frac{1}{2}\right)^{86} \cdot 3 \cdot \sqrt{2} \cdot \sin 18°(1 - \sin^2 18°) \\
&= \left(\frac{1}{2}\right)^{84} \cdot 3 \cdot \sqrt{2} \cdot \frac{\sqrt{5}-1}{4} \cdot \left(1 - \left(\frac{\sqrt{5}-1}{4}\right)^2\right) \\
&= \boxed{\frac{3}{2^{89}} \cdot \sqrt{10}}
\end{aligned}
$$

Chapter 8: Solutions

8.4 Trigonometric Techniques

Practice 1

Show that

$$\cos(\alpha) + \cos(\alpha+\beta) + \cdots + \cos(\alpha+n\beta) = \frac{\sin\frac{n+1}{2}\beta \cos\left(\alpha + \frac{n}{2}\beta\right)}{\sin\frac{\beta}{2}}$$

Multiplying each term of the left side with $2\sin\frac{\beta}{2}$ yields:

$$2\sin\frac{\beta}{2}\cos\alpha = \sin\left(\frac{\beta}{2}+\alpha\right) + \sin\left(\frac{\beta}{2}-\alpha\right)$$

$$2\sin\frac{\beta}{2}\cos(\alpha+\beta) = \sin\left(\frac{3\beta}{2}+\alpha\right) + \sin\left(-\frac{\beta}{2}-\alpha\right)$$

$$2\sin\frac{\beta}{2}\cos(\alpha+2\beta) = \sin\left(\frac{5\beta}{2}+\alpha\right) + \sin\left(-\frac{3\beta}{2}-\alpha\right)$$

$$\cdots = \cdots$$

$$2\sin\frac{\beta}{2}\cos(\alpha+n\beta) = \sin\left(\frac{(2n+1)\beta}{2}+\alpha\right) + \sin\left(-\frac{(2n-1)\beta}{2}-\alpha\right)$$

Adding these equations together and noting sine function is odd:

$$2\sin\frac{\beta}{2}(\cos(\alpha) + \cos(\alpha+\beta) + \cdots + \cos(\alpha+n\beta))$$
$$= \sin\left(\frac{2n+1}{2}\beta+\alpha\right) + \sin\left(\frac{\beta}{2}-\alpha\right)$$
$$= 2\sin\frac{n+1}{2}\beta \cos\left(\alpha + \frac{n}{2}\beta\right)$$

Dividing both sides by $2\sin\frac{\beta}{2}$ produces the desired result.

Chapter 8: Solutions

Practice 2

Show that

$$\cos\alpha + \cos 2\alpha + \cos 3\alpha + \cdots + \cos n\alpha = \frac{\sin\frac{n}{2}\alpha \cos\frac{n+1}{2}\alpha}{\sin\frac{\alpha}{2}}$$

Setting $\beta = \alpha$ in the previous practice immediately leads to the desired result. (Note there are $(n+1)$ terms in the previous practice and only n terms in this practice.)

Practice 3

Show that for any natural number n and real number $x \neq \frac{m\pi}{2^k}$ (where m in an integer), the following relation always holds:

$$\frac{1}{\sin 2x} + \frac{1}{\sin 4x} + \cdots + \frac{1}{\sin 2^n x} = \cot x - \cot 2^n x$$

The to-be-proved claim gives a strong hint that this left side can be transformed to a telescoping series. Furthermore, from the result, a natural guess is that

$$\frac{1}{\sin 2x} = \cot x - \cot 2x \quad , \quad \frac{1}{\sin 4x} = \cot 2x - \cot 4x, \cdots$$

Indeed, it is. Let's prove the first relation. All the others can be proved similarly.

$$\frac{1}{\sin 2x} = \frac{2\cos^2 x - \cos 2x}{\sin 2x} = \frac{2\cos^2 x}{2\sin x \cos x} - \frac{\cos 2x}{\sin 2x} = \cot x - \cot 2x$$

Note: this problem can also be solved using the complex method which will be discussed in *Chapter 5*.

Chapter 8: Solutions

Practice 4

Show that for any positive integer:

$$\tan x \tan 2x + \tan 2x \tan 3x + \cdots + \tan(n-1)x \tan nx = \frac{\tan nx}{\tan x} - n$$

Observing these terms finds that the difference of the two angles of each term equals x which is a constant. Hence, it is attempting to employ the difference of tangents formula *(3.3)* on *page 19*. A further hint is that there is a term of n in the right side of the target which can be sufficient to add 1 to each of these n terms on the left.

$$1 + \tan x \tan 2x = \frac{\tan 2x - \tan x}{\tan(2x - x)}$$

$$1 + \tan 2x \tan 3x = \frac{\tan 3x - \tan 2x}{\tan(3x - 2x)}$$

$$\cdots = \cdots$$

$$1 + \tan(n-1)x \tan nx = \frac{\tan nx - \tan(n-1)x}{\tan(nx - (n-1)x)}$$

Adding these $(n-1)$ equations together yields

$$(n-1) + \tan x \tan 2x + \tan 2x \tan 3x + \cdots + \tan(n-1)x \tan nx$$
$$= \frac{1}{\tan x}(\tan nx - \tan x)$$

Rearranging the last equation leads to the desired claim.

Practice 5

Show that

$$\tan x + 2 \tan 2x + 2^2 \tan 2^2 x + \cdots + 2^n \tan 2^n x = \cot x - 2^{n+1} \cot 2^{n+1} x$$

Similarly to the previous practice, the desired result suggests that the right side is a result of the following telescoping sequence:

$$(\cot x - 2 \cot 2x) + (2 \cot 2x - 2^2 \cot 2^2 x) + \cdots + (2^n \cot 2^n x - 2^{n+1} \cot 2^{n+1} x)$$

Let's evaluate each of these term:

$$\cot x - 2\cot 2x$$
$$= \frac{\cos x}{\sin x} - \frac{2\cos 2x}{\sin 2x}$$
$$= \frac{\cos x}{\sin x} - \frac{2(\cos^2 x - \sin^2 x)}{2\sin x \cos x}$$
$$= \frac{\cos^2 x}{\sin x \cos x} - \frac{\cos^2 x - \sin^2 x}{\sin x \cos x}$$
$$= \frac{\sin^2 x}{\sin x \cos x}$$
$$= \tan x$$

Therefore, we have

$$\tan x = \cot x - 2\cot 2x$$
$$2\tan 2x = 2\cot 2x - 2^2 \cot 2^2 x$$
$$\cdots = \cdots$$
$$2^n \tan 2^n x = 2^n \cot 2^n x - 2^{n+1} \cot 2^{n+1} x$$

Adding these n equations gives the desired result.

Practice 6

Show that
$$\cos\frac{\pi}{7} + \cos\frac{3\pi}{7} + \cos\frac{5\pi}{7} = \frac{1}{2}$$

This problem looks similar to *(4.2)* on *page 37* but with difference. It is possible to convert this problem to that of *(4.2)*. Noting that $\cos(\pi - \alpha) = -\cos\alpha$ yields:

$$\cos\frac{\pi}{7} + \cos\frac{3\pi}{7} + \cos\frac{5\pi}{7} = -\cos\frac{6\pi}{7} - \cos\frac{4\pi}{7} - \cos\frac{2\pi}{7}$$

Setting $n = 3$ in *(4.2)* gives

$$\cos\frac{2\pi}{7} + \cos\frac{4\pi}{7} + \cos\frac{6\pi}{7} = -\frac{1}{2}$$

Hence, the answer to this problem is $\boxed{\dfrac{1}{2}}$.

Practice 7

Compute the value of $(\cos\frac{\pi}{7} - \cos\frac{2\pi}{7} + \cos\frac{3\pi}{7})$.

This problem can also be solved using the similar way as the previous practice:

$$\cos\frac{\pi}{7} - \cos\frac{2\pi}{7} + \cos\frac{3\pi}{7}$$
$$= -\cos\frac{6\pi}{7} - \cos\frac{4\pi}{7} - \cos\frac{4\pi}{7}$$
$$= -\left(\cos\frac{2\pi}{7} + \cos\frac{4\pi}{7} + \cos\frac{6\pi}{7}\right)$$
$$= -\left(-\frac{1}{2}\right)$$
$$= \boxed{\dfrac{1}{2}}$$

Practice 8

Compute the value of $(\cos\frac{\pi}{9} + \cos\frac{3\pi}{9} + \cos\frac{5\pi}{9} + \cos\frac{7\pi}{9})$.

Setting $n = 4$ in (4.2) on *page 37* gives

$$\frac{2\pi}{9} + \frac{4\pi}{9} + \frac{6\pi}{9} + \frac{8\pi}{9} = -\frac{1}{2}$$

Therefore,

$$\cos\frac{\pi}{9} + \cos\frac{3\pi}{9} + \cos\frac{5\pi}{9} + \cos\frac{7\pi}{9}$$
$$= -\left(\cos\frac{8\pi}{9} + \cos\frac{6\pi}{9} + \cos\frac{4\pi}{9} + \cos\frac{2\pi}{9}\right)$$

$$= \boxed{\dfrac{1}{2}}$$

Practice 9

Show that

$$\frac{1}{\sin 1° \sin 2°} + \frac{1}{\sin 2° \sin 3°} + \cdots + \frac{1}{\sin 89° \sin 90°} = \cos 1° \csc^2 1°$$

This problem appears to be a fit for the telescoping technique. A natural guess is that

$$\frac{1}{sin 1° \sin 2°} = \frac{1}{K}\left(A - B\right)$$

where K is a constant and A and B are two trigonometric values which can form a telescoping sequence. In order to find this constant coefficient K, let's try to rewrite the right side to an appropriate form.

$$\frac{\cos 1°}{\sin^2 1°} = \frac{1}{\sin 1°} \cdot \cot 1°$$

Given the fact $\cot 90° = 0$, a conjuncture is that

$$\frac{1}{\sin 1° \sin 2°} + \cdots + \frac{1}{\sin 89° \sin 90°} = \frac{1}{\sin 1°}(\cot 1° - \cot 90°)$$

This is equivalent to showing that, for $n = 1, 2, \cdots, 89$,

$$\frac{1}{\sin n° \sin(n+1)°} = \frac{1}{\sin 1°} \cdot (\cot n° - \cot(n+1)°)$$

This indeed holds. Let's expanding the right side:

$$\frac{1}{\sin 1°} \cdot (\cot n° - \cot(n+1)°)$$
$$= \frac{1}{\sin 1°} \cdot \left(\frac{\cos n°}{\sin n°} - \frac{\cos(n+1)°}{\sin(n+1)°}\right)$$

Chapter 8: Solutions

$$= \frac{1}{\sin 1°} \cdot \frac{\cos n° \sin(n+1)° - \sin n° \cos(n+1)°}{\sin n° \sin(n+1)°}$$

$$= \frac{1}{\sin 1°} \cdot \frac{\sin 1°}{\sin n° \sin(n+1)°}$$

$$= \frac{1}{\sin n° \sin(n+1)°}$$

Practice 10

Compute the value of $(\sqrt{3}\tan 18° + \tan 18° \tan 12° + \sqrt{3}\tan 12°)$.

Noting that $18° + 12° = 30°$ which is a special angle reminds the sum of tangent formula.

$$\sqrt{3}\tan 18° + \tan 18° \tan 12° + \sqrt{3}\tan 12°$$
$$= \sqrt{3}\cdot(\tan 18° + \tan 12°) + \tan 18° \tan 12°$$
$$= \sqrt{3}\cdot \tan(18° + 12°)(1 - \tan 18° \tan 12°) + \tan 18° \tan 12°$$
$$= \sqrt{3}\cdot \frac{\sqrt{3}}{3}\cdot(1 - \tan 18° \tan 12°) + \tan 18° \tan 12°$$
$$= \boxed{1}$$

Practice 11

Compute

$$\cos\frac{\pi}{2n+1}\cdot \cos\frac{2\pi}{2n+1}\cdots\cos\frac{n\pi}{2n+1}$$

Let

$$C = \cos\frac{\pi}{2n+1}\cdot \cos\frac{2\pi}{2n+1}\cdots\cos\frac{n\pi}{2n+1}$$

and

$$S = \sin\frac{\pi}{2n+1}\cdot \sin\frac{2\pi}{2n+1}\cdots\sin\frac{n\pi}{2n+1}$$

Then,

$$C \cdot S = \left(\cos\frac{\pi}{2n+1} \cdot \cos\frac{2\pi}{2n+1} \cdots \cos\frac{n\pi}{2n+1}\right)$$
$$\left(\sin\frac{\pi}{2n+1} \cdot \sin\frac{2\pi}{2n+1} \cdots \sin\frac{n\pi}{2n+1}\right)$$
$$= \left(\sin\frac{\pi}{2n+1}\cos\frac{\pi}{2n+1}\right)\left(\sin\frac{2\pi}{2n+1}\cos\frac{2\pi}{2n+1}\right)$$
$$\cdots\left(\sin\frac{n\pi}{2n+1}\cos\frac{n\pi}{2n+1}\right)$$
$$= \left(\frac{1}{2}\cdot\sin\frac{2\pi}{2n+1}\right)\left(\frac{1}{2}\cdot\sin\frac{4\pi}{2n+1}\right)$$
$$\cdots\left(\frac{1}{2}\cdot\sin\frac{2n-2\pi}{2n+1}\right)\left(\frac{1}{2}\cdot\sin\frac{2n\pi}{2n+1}\right)$$
$$= \left(\frac{1}{2}\cdot\sin\frac{2\pi}{2n+1}\right)\left(\frac{1}{2}\cdot\sin\frac{4\pi}{2n+1}\right)$$
$$\cdots\left(\frac{1}{2}\cdot\sin\frac{3\pi}{2n+1}\right)\left(\frac{1}{2}\cdot\sin\frac{\pi}{2n+1}\right)$$
$$= \frac{1}{2^n}\cdot S$$

$$\therefore\ C = \boxed{\frac{1}{2^n}}$$

Practice 12

Prove that $\cos 1°$ is irrational.

Let's prove this using proof by contradiction and mathematical induction[1].

Assuming $\cos 1°$ is rational. Then $\cos 2°$ will be rational too because $\cos 2° = 2\cos^2 1° - 1$. Meanwhile, if both $\cos n°$ and $\cos(n-1)°$ are rational where n is a positive integer greater than 1, so will

---

[1] Both methods are discussed in the book *Art of Thinking* written by the same author.

Chapter 8: Solutions

be $\cos(n° + 1°)$ because
$$\cos(n° + 1°) + \cos(n° - 1°) = 2\cos n° \cos 1°$$

Therefore, by the principle of mathematical induction, $\cos n°$ is rational for any positive integer n.

However, this conclusion is clearly false because $\cos 30°$ is irrational. This follows that $\cos 1°$ cannot be rational, i.e., it must be irrational.

Practice 13

Find the minimal value of
$$\left|\sin x + \cos x + \tan x + \cot x + \sec x + \csc x\right|$$
where x is a real number.

(Putnam)

First, let's simplify the given expression to contain just $\sin x$ and $\cos x$:

$$f(x) = \left|\sin x + \cos x + \tan x + \cot x + \sec x + \csc x\right|$$
$$= \left|\sin x + \cos x + \frac{\sin x}{\cos x} + \frac{\cos x}{\sin x} + \frac{1}{\cos x} + \frac{1}{\sin x}\right|$$
$$= \left|\frac{\sin^2 x \cos x + \cos^2 x \sin x + \sin^2 x + \cos^2 x + \sin x + \cos x}{\sin x \cos x}\right|$$
$$= \left|\frac{\sin x \cos x(\sin x + \cos x) + 1 + (\sin x + \cos x)}{\sin x \cos x}\right|$$
$$= \left|(\sin x + \cos x) + \frac{1 + (\sin x + \cos x)}{\sin x \cos x}\right|$$

Let $y = \sin x + \cos x$. Then
$$y^2 = 1 + 2\sin x \cos x \implies \sin x \cos x = \frac{1}{2}(y^2 - 1)$$

Therefore $f(x)$ is equivalent to
$$f(y) = \left|y + \frac{2(1 + y)}{y^2 - 1}\right| = \left|y + \frac{2}{y - 1}\right| = \left|(y - 1) + \frac{2}{y - 1} + 1\right|$$

116

The range of y is $[-\sqrt{2}, \sqrt{2}]$ because

$$y = \sin x + \cos x = \sqrt{2}\sin(x + 45°)$$

i) If $(y-1) > 0$, or equivalently, $y \in (1, \sqrt{2}]$, then

$$(y-1) + \frac{2}{y-1} + 1 \geq 2\sqrt{(y-1) \cdot \frac{2}{y-1}} + 1 = 2\sqrt{2} + 1 \geq 0$$

$$\implies f(y) \geq 2\sqrt{2} + 1$$

ii) If $(y-1) < 0$, or equivalently, $y \in [-\sqrt{2}, 1)$, then

$$(y-1) + \frac{2}{y-1} + 1 \geq -2\sqrt{(1-y) \cdot \frac{2}{1-y}} + 1 = -2\sqrt{2} + 1 < 0$$

$$f(y) \geq 2\sqrt{2} - 1$$

Therefore, we conclude that the minimal value of given expression equals $\boxed{2\sqrt{2} - 1}$ which is reachable when

$$1 - y = \frac{2}{1-y}, y < 1 \implies y = 1 - \sqrt{2} \implies \sin(x + 45°) = \frac{\sqrt{2}}{2} - 1$$

Chapter 8: Solutions

8.5 The Complex Number Method

Practice 1

Prove the following formula using the complex number method:
$$\sin(\alpha + \beta) = \sin\alpha\cos\beta + \cos\alpha\sin\beta$$
$$\cos(\alpha + \beta) = \cos\alpha\cos\beta - \sin\alpha\sin\beta$$

Let $z_1 = \cos\alpha + i\sin\alpha$ and $z_2 = \cos\beta + i\sin\beta$. Then, on one hand, we have
$$z_1 z_2 = \cos(\alpha+\beta) + i\sin(\alpha+\beta)$$
And, on the other hand,
$$\begin{aligned} z_1 z_2 &= (\cos\alpha + i\sin\alpha)(\cos\beta + i\sin\beta) \\ &= (\cos\alpha\cos\beta - \sin\alpha\sin\beta) + i(\sin\alpha\cos\beta + \cos\alpha\sin\beta) \end{aligned}$$

These two results must agree with each other. Matching their real and imaginary, respectively, gives the desired result.

Practice 2

Solve the equation $\cos^2 x + \cos^2 2x + \cos^2 3x = 1$ in $(0, 2\pi)$.
(1962 IMO)

The given equation can be rewritten as
$$\frac{1+\cos 2x}{2} + \frac{1+\cos 4x}{2} + \frac{1+\cos 6x}{2} = 1$$
$$\implies \cos 2x + \cos 4x + \cos 6x = -1$$
Let $z = \cos x + i\sin x$. Then
$$z^2 + \frac{1}{z^2} + z^4 + \frac{1}{z^4} + z^6 + \frac{1}{z^6} = -2$$

Let $w = z^2 + \frac{1}{z^2}$, then

$$z^4 + \frac{1}{z^4} = \left(z^2 + \frac{1}{z^2}\right)^2 - 2 = w^2 - 2$$

$$z^6 + \frac{1}{z^6} = \left(z^2 + \frac{1}{z^2}\right)^3 - 3 \cdot z^2 \cdot \frac{1}{z^2}\left(z^2 + \frac{1}{z^2}\right) = w^3 - 3w$$

Therefore, the original equation is equivalent to

$$w^3 + w^2 - 2w = 0 \implies w(w+2)(w-1) = 0 \implies w = 0, 1, -2$$

When $w = 0$: $z^2 + \frac{1}{z^2} = 0 \implies z^4 = -1$

$$\therefore x = 45°, 135°, 225°, 315°$$

When $w = 1$: $z^2 + \frac{1}{z^2} = 1 \implies z^4 - z^2 + 1 = 0 \implies z^2 = \cos 60° + i \sin 60°$, or $\cos 300° + i \sin 300°$

$$\therefore x = 30°, 150°, 210°, 330°$$

When $w = -2$: $z^2 + \frac{1}{z^2} = -2 \implies (z^2 - 1)^2 = 0 \implies z = \pm 1$

$$\therefore x = 180°$$

Hence, we conclude

$$x = \boxed{30°, 45°, 135°, 150°, 180°, 210°, 225°, 315°, 330°}$$

Practice 3

Prove the following identities

$$\sin(3\theta) = 3\sin\theta - 4\sin^3\theta$$
$$\cos(3\theta) = 4\cos^3\theta - 3\cos\theta$$

Let $z = \cos\alpha + i\sin\alpha$. Then we have

$$\cos 3\theta + i \sin 3\theta = (\cos\theta + i\sin\theta)^3 \tag{8.7}$$

Chapter 8: Solutions

Expanding the right side of *(8.7)* using binomial formula yields:

$(\cos\theta + i\sin\theta)^3$
$= \cos^3\theta + 3\cos^2\theta(i\sin\theta) + 3\cos\theta(i\sin\theta)^2 + (i\sin\theta)^3$
$= (\cos^3\theta - 3\cos\theta\sin^2\theta) + i(3\cos^2\theta\sin\theta - \sin^3\theta)$
$= (\cos^3\theta - 3\cos\theta(1 - \cos^2\theta)) + i(3(1 - \sin^2\theta)\sin\theta - \sin^3\theta)$
$= (4\cos^3\theta - 3\cos\theta) + i(3\sin\theta - 4\sin^3\theta)$

$\therefore \quad \cos 3\theta + i\sin 3\theta = (4\cos^3\theta - 3\cos\theta) + i(3\sin\theta - 4\sin^3\theta)$

Matching the real and imaginary parts of this equation leads to the result immediately.

Practice 4

Solve the equation $\cos\theta + \cos 2\theta + \cos 3\theta = \sin\theta + \sin 2\theta + \sin 3\theta$.

Let $z = \cos\theta + i\sin\theta$, then the give equation is equivalent to

$$\frac{1}{2} \times \left(\left(z + \frac{1}{z}\right) + \left(z^2 + \frac{1}{z^2}\right) + \left(z^3 + \frac{1}{z^3}\right)\right)$$
$$= -\frac{i}{2} \times \left(\left(z - \frac{1}{z}\right) + \left(z^2 - \frac{1}{z^2}\right) + \left(z^3 - \frac{1}{z^3}\right)\right)$$

or

$(z^6 + z^5 + z^4)(1 + i) + (z^2 + z + 1)(1 - i) = 0$
$(z^2 + z + 1)(z^4(1 + i) + (1 - i)) = 0$
$(z^2 + z + 1)(z^4 - i) = 0$

Therefore

$z^2 + z + 1 = 0 \implies \theta = \boxed{\left((2k + 1) \pm \frac{1}{3}\right)\pi}$

$z^4 - i = 0 \implies \theta = \boxed{\left(\frac{k}{2} + \frac{1}{8}\right)\pi}$

where k is an integer.

Practice 5

If $\sin A + \sin B + \sin C = 0 = \cos A + \cos B + \cos C = 0$, explain for any positive integer n, it must hold that

$$\sin nA + \sin nB + \sin nC = \cos nA + \cos nB + \cos nC = 0$$

Let $z_a = \cos A + i \sin A$, $z_b = \cos B + i \sin B$, and $z_c = \cos C + i \sin C$, then $|z_a| = |z_b| = |z_c| = 1$ and

$$z_a + z_b + z_c = (\cos A + \cos B + \cos C) + i(\sin A + \sin B + \sin C) = 0$$

This means that the sum of three unit vectors equals 0. This can only happen when these three vectors are sides of an equilateral triangle. Or in another word, A, B, C equal (in some order of) θ, $\theta + 120°$ and $\theta - 120°$.

Consequently, nA, nB and nC will be equal to $n\theta$, $n(\theta - 120°)$ and $n(\theta + 120°)$. The sum of their corresponding vectors must equal 0 too. This is equivalent to the to-be-proved claim because the sum of these three new vectors can be written as

$$(\cos nA + \cos nB + \cos nC) + i(\sin nA + \sin nB + \sin nC) = 0$$

Practice 6

Prove: $\cos 7x + 7 \cos 5x + 21 \cos 3x + 35 \cos x = 64 \cos^7 x$.

Let $z = \cos x + i \sin x$, then $\cos x = \frac{1}{2}\left(z + \frac{1}{z}\right)$. It follows that

$64 \cos^7 x$

$$= 64 \cdot \left(\frac{1}{2} \cdot \left(z + \frac{1}{z}\right)\right)^7$$

$$= \frac{1}{2} \cdot \left(z^7 + 7z^5 + 21z^3 + 35z + 35\frac{1}{z} + 21\frac{1}{z^3} + 7\frac{1}{z^5} + \frac{1}{z^7}\right)$$

$$= \frac{1}{2} \cdot \left(\left(z^7 + \frac{1}{z^7}\right) + 7 \cdot \left(z^5 + \frac{1}{z^5}\right) + 21\left(z^3 + \frac{1}{z^3}\right) + 35 \cdot \left(z + \frac{1}{z}\right)\right)$$

Chapter 8: Solutions

$$= \frac{1}{2} \cdot 2 \cdot \left(\cos 7x + 7\cos 5x + 21\cos 3x + 35\cos x \right)$$
$$= \cos 7x + 7\cos 5x + 21\cos 3x + 35\cos x$$

Practice 7

Let $A(x_1, y_1)$, $B(x_2, y_2)$, and $C(x_3, y_3)$ be three points on the unit circle, and

$$x_1 + x_2 + x_3 = y_1 + y_2 + y_3 = 0$$

Prove
$$x_1^2 + x_2^2 + x_3^2 = y_1^2 + y_2^2 + y_3^2 = \frac{3}{2}$$

Given $\overrightarrow{OA} + \overrightarrow{OB} + \overrightarrow{OC} = 0$, it is easy to see that

$$\angle AOB = \angle BOC = \angle COA = 120°$$

Let $A = (\cos \alpha, \sin \alpha)$. Then $B = (\cos(\alpha + 120°), \sin(\alpha + 120°))$ and $C = (\cos(\alpha - 120°), \sin(\alpha - 120°))$. Then,

$$x_1^2 + x_2^2 + x_3^2$$
$$= \cos^2 \alpha + \cos^2(\alpha + 120°) + \cos^2(\alpha - 120°)$$
$$= \frac{1 + \cos 2\alpha}{2} + \frac{1 + \cos(2\alpha + 240°)}{2} + \frac{1 + \cos(2\alpha - 240°)}{2}$$
$$= \frac{3}{2} + \frac{1}{2} \cdot (\cos 2\alpha + \cos(2\alpha + 240°) + \cos(2\alpha - 240°))$$
$$= \frac{3}{2} + \frac{1}{2} \cdot (\cos 2\alpha + 2\cos 2\alpha \cos 240°)$$
$$= \frac{3}{2}$$

Meanwhile, it is clear that

$$x_1^2 + x_2^2 + x_3^2 + y_1^2 + y_2^2 + y_3^2$$
$$= (x_1^2 + y_1^2) + (x_2^2 + y_2^2) + (x_3^2 + y_3^2)$$

$$= (\cos^2 \alpha + \sin^2 \alpha) + (\cos^2(\alpha + 120°) + \sin^2(\alpha + 120°))$$
$$+ (\cos^2(\alpha - 120°) + \sin^2(\alpha - 120°))$$
$$= 3$$

Therefore $y_1^2 + y_2^2 + y_3^2 = \frac{3}{2}$ too.

Practice 8

Compute the values of
$$S = C_n^1 \sin \theta + C_n^2 \sin 2\theta + \cdots + C_n^n \sin n\theta$$
and
$$C = C_n^1 \cos \theta + C_n^2 \cos 2\theta + \cdots + C_n^n \cos n\theta$$

Let $z = \cos \theta + i \sin \theta$. Then we have $z^n = \cos n\theta + i \sin n\theta$.

$\therefore 1 + C + iS$
$$= 1 + C_n^1(\cos \theta + i \sin \theta) + C_n^2(\cos 2\theta + i \sin 2\theta) + \cdots$$
$$+ C_n^n(\cos n\theta + i \sin n\theta)$$
$$= C_n^0 z^0 + C_n^1 z^1 + C_n^2 z^2 + \cdots + C_n^n z^n$$
$$= (1 + z)^n$$

Meanwhile, we have
$$(1+z)^n = (1 + \cos\theta + i \sin \theta)^n$$
$$= \left(2\cos^2 \frac{\theta}{2} + 2\sin \frac{\theta}{2} \cos \frac{\theta}{2} i\right)^n$$
$$= 2^n \cos^n \frac{\theta}{2} \left(\cos \frac{\theta}{2} + i \sin \frac{\theta}{2}\right)^n$$
$$= 2^n \cos^n \frac{\theta}{2} \left(\cos \frac{n\theta}{2} + i \sin \frac{n\theta}{2}\right)$$

Therefore, we conclude
$$S = \boxed{2^n \cos^n \frac{\theta}{2} \sin \frac{n\theta}{2}} \quad \text{and} \quad C = \boxed{-1 + 2^n \cos^n \frac{\theta}{2} \cos \frac{n\theta}{2}}$$

Chapter 8: Solutions

8.6 Trigonometry in Triangle

Practice 1

In $\triangle ABC$, if $\frac{a}{b} = 2 + \sqrt{3}$ and $\angle C = 60°$, find the measurement of $\angle A$ and $\angle B$.

Because $A + B + C = 180°$, we have $A + B = 120°$. Therefore,
$$\tan \frac{A+B}{2} = \tan 60° = \sqrt{3}$$

Meanwhile,
$$\frac{a+b}{a-b} = \frac{\frac{a}{b}+1}{\frac{a}{b}-1} = \frac{2+\sqrt{3}+1}{2+\sqrt{3}-1} = \sqrt{3}$$

Applying the Law of Tangents *(6.6)* on *page 63*:
$$\frac{a+b}{a-b} = \frac{\tan \frac{A+B}{2}}{\tan \frac{A-B}{2}} \implies \tan \frac{A-B}{2} = 1 \implies A - B = 90°$$

Considering this and $A + B = 120°$ gives
$$A = \boxed{105°} \quad \text{and} \quad B = \boxed{15°}$$

Practice 2

Given $\triangle ABC$, show that
$$\frac{a}{b+c} \geq \sin \frac{A}{2}$$

By the Law of Sines and the sum-to-product formula:
$$\frac{a}{b+c} = \frac{\sin A}{\sin B + \sin C} = \frac{2 \sin \frac{A}{2} \cos \frac{A}{2}}{2 \sin \frac{B+C}{2} \cos \frac{B-C}{2}}$$

Because $A + B + C = 180°$, therefore
$$\sin\frac{B+C}{2} = \sin\left(90° - \frac{A}{2}\right) = \cos\frac{A}{2}$$

Setting this to the previous relation leads to
$$\frac{a}{b+c} = \frac{\sin\frac{A}{2}}{\cos\frac{B-C}{2}} \geq \sin\frac{A}{2}$$

because $\cos\frac{B-C}{2} \leq 1$.

Practice 3

Given $\triangle ABC$, show that
$$\cos A + \cos B + \cos C = 1 + 4\sin\frac{A}{2}\sin\frac{B}{2}\sin\frac{C}{2}$$

This identity can be proved directly as follow:
$$\cos A + \cos B + \cos C$$
$$= (\cos A + \cos B) - \cos(A+B)$$
$$= 2\cos\frac{A+B}{2}\cos\frac{A-B}{2} - \left(2\cos^2\frac{A+B}{2} - 1\right)$$
$$= 1 + 2\cos\frac{A+B}{2}\left(\cos\frac{A-B}{2} - \cos\frac{A+B}{2}\right)$$
$$= 1 + 2\sin\frac{C}{2}\cdot\left(2\sin\frac{A}{2}\sin\frac{B}{2}\right)$$
$$= 1 + 4\sin\frac{A}{2}\sin\frac{B}{2}\sin\frac{C}{2}$$

Practice 4

In $\triangle ABC$, show that
$$\cot A \cot B + \cot B \cot C + \cot C \cot A = 1$$

Chapter 8: Solutions

If $\triangle ABC$ is right, without loss of generality, let's assume $A = 90°$. Then $\cot A = 0$ and $\cot B = \tan C$ because $B + C = 90°$. Then

$$\cot A \cot B + \cot B \cot C + \cot C \cot A = 0 + \cot B \cot C + 0 = 1$$

Otherwise, if none of the three angles is right, then all of $\tan A$, $\tan B$ and $\tan C$ are well defined. Multiplying both sides of the to-be-claimed relations by $\tan A \tan B \tan C$ yields

$$\tan A + \tan B + \tan C = \tan A \tan B \tan C$$

This is known to hold by *(6.10)* on *page 67*.

Therefore, we conclude the relation always holds.

Practice 5

Show that

$$\tan\frac{A}{2}\tan\frac{B}{2} + \tan\frac{B}{2}\tan\frac{C}{2} + \tan\frac{C}{2}\tan\frac{A}{2} = 1$$

By *(3.5)* on *page 22*, we have

$$\tan\frac{A}{2}\tan\frac{B}{2} + \tan\frac{B}{2}\tan\frac{C}{2} + \tan\frac{C}{2}\tan\frac{A}{2}$$
$$= \tan\frac{A}{2}\tan\frac{B}{2} + \tan\frac{C}{2}\left(\tan\frac{A}{2} + \tan\frac{B}{2}\right)$$
$$= \tan\frac{A}{2}\tan\frac{B}{2} + \tan\frac{C}{2}\tan\left(\frac{A}{2} + \frac{B}{2}\right)\left(1 - \tan\frac{A}{2}\tan\frac{B}{2}\right)$$

Meanwhile, we have

$$A+B+C = \pi \implies \frac{A}{2}+\frac{B}{2} = \frac{\pi}{2}-\frac{C}{2} \implies \tan\frac{C}{2}\tan\left(\frac{A}{2}+\frac{B}{2}\right) = 1$$

Setting this to the previous relation yields the desired result immediately.

Practice 6

In $\triangle ABC$, show that

$$\tan\frac{A}{2}\tan\frac{B}{2}\tan\frac{C}{2} \le \frac{\sqrt{3}}{9}$$

Applying the AM-GM inequality on the conclusion of the previous practice:

$$1 = \tan\frac{A}{2}\tan\frac{B}{2} + \tan\frac{B}{2}\tan\frac{C}{2} + \tan\frac{C}{2}\tan\frac{A}{2}$$

$$\ge 3\sqrt[3]{\left(\tan\frac{A}{2}\tan\frac{B}{2}\right)\left(\tan\frac{B}{2}\tan\frac{C}{2}\right)\left(\tan\frac{C}{2}\tan\frac{A}{2}\right)}$$

$$= 3\sqrt[3]{\left(\tan\frac{A}{2}\tan\frac{B}{2}\tan\frac{C}{2}\right)^2}$$

$$\therefore \quad \tan\frac{A}{2}\tan\frac{B}{2}\tan\frac{C}{2} \le \frac{\sqrt{3}}{9}$$

Practice 7

In $\triangle ABC$, show that

$$\sin 2A + \sin 2B + \sin 2C = 4\sin A \sin B \sin C$$
$$\cos 2A + \cos 2B + \cos 2C = -1 - 4\cos A \cos B \cos C$$

Both identities can be proved by applying basic formulas.

$$\sin 2A + \sin 2B + \sin 2C$$
$$= 2\sin(A+B)\cos(A-B) + 2\sin C \cos C$$
$$= 2\sin C \cos(A-B) + 2\sin C \cos C$$
$$= 2\sin C \cdot (\cos(A-B) - \cos(A+B))$$
$$= 2\sin C \cdot 2\sin A \sin B$$
$$= 4\sin A \sin B \sin C$$

Chapter 8: Solutions

$$\cos 2A + \cos 2B + \cos 2C$$
$$= 2\cos(A+B)\cos(A-B) + 2\cos^2 C - 1$$
$$= -2\cos C \cos(A-B) - 2\cos C \cos(A+B) - 1$$
$$= -2\cos C(\cos(A-B) + \cos(A+B)) - 1$$
$$= -4\cos A \cos B \cos C - 1$$

Practice 8

In $\triangle ABC$, show that

$$\sin^2 A + \sin^2 B + \sin^2 C = 2 + 2\cos A \cos B \cos C$$
$$\cos^2 A + \cos^2 B + \cos^2 C = 1 - 2\cos A \cos B \cos C$$

The first equation can be derived from the previous practice.

$$\sin^2 A + \sin^2 B + \sin^2 C$$
$$= \frac{1 - \cos 2A}{2} + \frac{1 - \cos 2B}{2} + \frac{1 - \cos 2C}{2}$$
$$= \frac{1}{2} \cdot (3 - (\cos 2A + \cos 2B + \cos 2C))$$
$$= \frac{1}{2} \cdot (3 - (-1 - 4\cos A \cos B \cos C))$$
$$= 2 + 2\cos A \cos B \cos C$$

Meanwhile,

$$\cos^2 A + \cos^2 B + \cos^2 C$$
$$= 3 - (\sin^2 A + \sin^2 B + \sin^2 C)$$
$$= 3 - (2 + 2\cos A \cos B \cos C)$$
$$= 1 - 2\cos A \cos B \cos C$$

Chapter 8: Solutions

Practice 9

Given $\triangle ABC$, show that

$$\cos A \cos B \cos C \leq \frac{1}{8}$$

$$\cos^2 A + \cos^2 B + \cos^2 C \geq \frac{3}{4}$$

If $\triangle ABC$ is non-acute, then one of $\cos A$, $\cos B$, and $\cos C$ is non-positive which means

$$\cos A \cos B \cos C \leq 0 < \frac{1}{8}$$

If $\triangle ABC$ is acute, then all of these three terms are positive. From the conclusion of the previous practice, we have

$$\cos^2 A + \cos^2 B + \cos^2 C = 1 - 2\cos A \cos B \cos C$$

$$\begin{aligned}
\therefore \quad 1 - 2\cos A \cos B \cos C &\\
= \cos^2 A + \cos^2 B + \cos^2 C &\\
\geq 3\sqrt[3]{\cos^2 A \cos^2 B \cos^2 C} &
\end{aligned}$$

Let $t = \sqrt[3]{\cos A \cos B \cos C}$. Then the previous inequality means

$$1 - 2t^3 \geq 3t^2 \Leftrightarrow (t+1)^2(2t-1) \leq 0 \Leftrightarrow t \leq \frac{1}{2}$$

$$\therefore \quad \cos A \cos B \cos C = t^3 \leq \frac{1}{8}$$

Then,

$$\cos^2 A + \cos^2 B + \cos^2 = 1 - 2\cos A \cos B \cos C \geq 1 - 2 \times \frac{1}{8} = \frac{3}{4}$$

Practice 10

In $\triangle ABC$, show that

$$2R \sin A \sin B \sin C = r(\sin A + \sin B + \sin C)$$

where R is the circumradius and r is the inradius.

The to-be-proved claim is equivalent to the following by multiplying both sides by R:

$$2R^2 \sin A \sin B \sin C = rR(\sin A + \sin B + \sin C)$$

Now, the left side equals the area of $\triangle ABC$ by *(6.7)* on *page 64*. The right side also represents the area of $\triangle ABC$ by first applying the law of sines and then the $S = rp$ formula:

$$rR(\sin A + \sin B + \sin C) = r \cdot \frac{1}{2}(a + b + c) = S_{\triangle ABC}$$

Practice 11

In $\triangle ABC$, $\angle C = \angle A + 60°$. If $BC = 1$, $AC = r$ and $AB = r^2$, where $r > 1$, prove $r \leq \sqrt{2}$.

Let $\angle A = \alpha$, then $\angle C = 60° + \alpha$ and $\angle B = 120° - 2\alpha$.

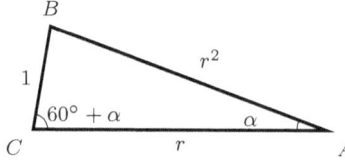

By the Law of Cosines:

$$r^2 = 1^2 + (r^2)^2 - 2 \cdot 1 \cdot r^2 \cdot \cos(120° - 2\alpha)$$

Chapter 8: Solutions

or
$$r^4 + 1 = r^2 + 2 \cdot r^2 \cdot \cos(120° - 2\alpha) \qquad (8.8)$$

Meanwhile, because $r^4 + 1 \geq 2r^2$, therefore,

$$r^2 + 2 \cdot r^2 \cdot \cos(120° - 2\alpha) \geq 2r^2$$
$$\Longrightarrow \quad \cos(120° - 2\alpha) \geq \frac{1}{2}$$
$$\Longrightarrow \quad 60° > \alpha \geq 30°$$

By the Law of Sines:

$$\begin{aligned}
\frac{r^2}{1} &= \frac{\sin(60° + \alpha)}{\sin \alpha} \\
&= \frac{\sin 60° \cos \alpha + \cos 60° \sin \alpha}{\sin \alpha} \\
&= \frac{\sqrt{3}}{2} \cdot \cot \alpha + \frac{1}{2} \\
&\leq \frac{\sqrt{3}}{2} \cdot \sqrt{3} + \frac{1}{2} \\
&= 2
\end{aligned}$$

$$\therefore \quad r \leq \sqrt{2}$$

Practice 12

Prove that there is one and only one triangle whose side lengths are consecutive integers, and one of whose angles is twice as large as another.
(IMO 1968)

Let $A = \alpha$ and $B = 2A = 2\alpha$. Then by the Law of Sines:

$$\frac{a}{\sin \alpha} = \frac{b}{\sin 2\alpha} \Longrightarrow \frac{a}{\sin \alpha} = \frac{b}{2 \sin \alpha \cos \alpha} \Longrightarrow \cos \alpha = \frac{b}{2a}$$

Chapter 8: Solutions

Meanwhile, by Law of Cosines *(6.4)* on *page 62*,

$$\cos\alpha = \cos A = \frac{b^2 + c^2 - a^2}{2bc} \implies \frac{b}{2a} = \frac{b^2 + c^2 - a^2}{2bc}$$

Rearranging and factorizing this relation give

$$b^2 c = a(b^2 + c^2 - a^2)$$
$$b^2 c = a(b^2 - a^2 - ac) + ac^2 + a^2 c$$
$$b^2 c - ac^2 - a^2 c = a(b^2 - a^2 - ac)$$
$$c(b^2 - ac - a^2) = a(b^2 - a^2 - ac)$$
$$(c - a)(b^2 - ac - a^2) = 0$$

Because a, b, and c are consecutive integers, $c \neq a$. This means that

$$b^2 - ac - a^2 = 0 \tag{8.9}$$

Meanwhile, because $B > A$, it must have $b > a$. It follows that there are three possible scenarios:

i) $c > b > a \implies c = a + 2, b = a + 1$

ii) $b > c > a \implies b = a + 2, c = a + 1$

iii) $b > a > c \implies b = a + 1, c = a - 1$

When $c = a + 2$ and $b = a + 1$, then *(8.9)* means

$$(a+1)^2 - a(a+2) - a^2 = 0 \implies a = \pm 1$$

Discarding negative value gives $a = 1$ which implies $b = 2$ and $c = 3$. However, in this case, $\triangle ABC$ will become degenerated.

When $b = a + 2$ and $c = a + 1$, then *(8.9)* means

$$(a+2)^2 - a(a+1) - a^2 = 0 \implies a = -1, 4$$

Discarding negative value gives $a = 4$ which gives $b = 6$ and $c = 5$.

When $b = a+1$ and $c = a-1$, then *(8.9)* means
$$(a+1)^2 - a(a-1) - a^2 = 0 \implies a = \frac{1}{2} \cdot (3 \pm \sqrt{13})$$
Neither of them is integer.

Therefore, there is only one triangle satisfying the requirement. The three sides of this triangle is 4, 5 and 6.

Alternative Solution

This problem can also be solved without using trigonometry.

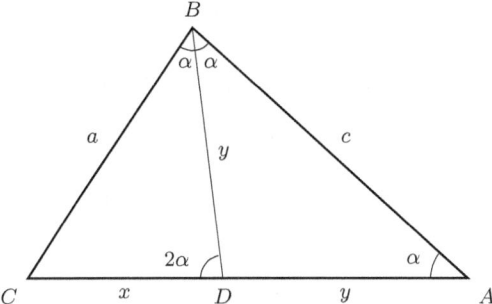

As shown above, in $\triangle ABC$, let $\angle B = 2\angle A$ and BD be the angle bisector of $\angle B$. Note that $x + y = b$. Then
$$\triangle ABC \sim \triangle BDC \implies \frac{AB}{BD} = \frac{BC}{DC} = \frac{AC}{BC}$$
$$\therefore \quad \frac{c}{y} = \frac{a}{x} = \frac{x+y}{a}$$
$$\implies \quad ac = y(x+y) \quad \text{and} \quad a^2 = x(x+y)$$

Adding these two relations and replace $x + y$ with b leads
$$ac + a^2 = b^2$$

Then, this equation can be analyzed using the same casework technique as the one employed in the original solution to find the final result.

Chapter 8: Solutions

8.7 Additional Techniques

Practice 1

Let real numbers x and y satisfy the relation $4x^2 - 5xy + 4y^2 = 5$. Find the maximum and minimal value of $x^2 + y^2$.

This problem can be solved using AM-GM inequality or the geometry method. Here, we present a trigonometry based solution.

Let $r = x^2 + y^2$. Then there exists θ so that $x = \sqrt{r}\cos\theta$ and $y = \sqrt{r}\sin\theta$. Setting these to the given relation leads to

$$4r\cos^2\theta - 5r\sin\theta\cos\theta + 4r\sin^2\theta = 5$$
$$4r(\cos^2\theta + \sin^2\theta) - 5r\sin\theta\cos\theta = 5$$
$$4r - \frac{5}{2}r\sin 2\theta = 5$$
$$\therefore \quad r = \frac{10}{8 - 5\sin 2\theta}$$

Now, it is clear that when $\sin 2\theta = 1$, r has a maximum value of $\boxed{\dfrac{10}{3}}$ and when $\sin 2\theta = -1$, r reaches minimal at $\boxed{\dfrac{10}{13}}$.

Practice 2

Given non-negative real numbers x, y and z, prove

$$\sqrt{x^2 + y^2 - xy} + \sqrt{y^2 + z^2 - yz} \geq \sqrt{x^2 + z^2 + xz}$$

This problem can be solved geometrically by constructing the following graph.

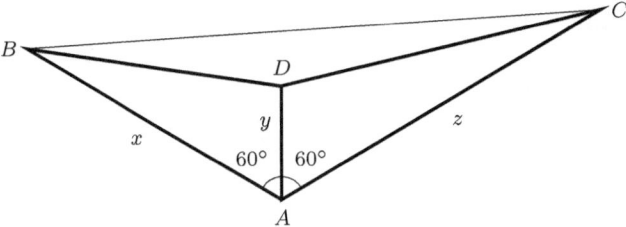

By the law of cosines, we find

$$BD = \sqrt{x^2 + y^2 - xy}$$
$$CD = \sqrt{y^2 + z^2 - yz}$$
$$BC = \sqrt{x^2 + z^2 + xz}$$

Noting $BD + CD \geq BC$ in $\triangle BCD$ leads to the conclusion immediately. The equality holds if and only if BCD are collinear.

Practice 3

Solve this inequality

$$\frac{x}{\sqrt{x^2+1}} + \frac{1-x^2}{1+x^2} > 0$$

Let $x = \tan\theta$ where $-\frac{\pi}{2} < \theta < \frac{\pi}{2}$, then

$$\begin{aligned}
&& \frac{x}{\sqrt{x^2+1}} + \frac{1-x^2}{1+x^2} &> 0 \\
\Leftrightarrow && \frac{\tan\theta}{\sqrt{\tan^2\theta+1}} + \frac{1-\tan^2\theta}{1+\tan^2\theta} &> 0 \\
\Leftrightarrow && \sin\theta + (\cos^2\theta - \sin^2\theta) &> 0 \\
\Leftrightarrow && \sin\theta + (1 - 2\sin^2\theta) &> 0 \\
\Leftrightarrow && (2\sin\theta + 1)(\sin\theta - 1) &< 0
\end{aligned}$$

Because $(\sin\theta - 1) \leq 0$ always hold unless $\sin\theta = 1$, the above relation can hold if and only if

$2\sin\theta+1 > 0$ and $\sin\theta < 1 \implies 1 > \sin\theta > -\frac{1}{2} \implies \frac{\pi}{2} > \theta > -\frac{\pi}{6}$

Chapter 8: Solutions

Accordingly, we find

$$\boxed{x > -\frac{\sqrt{3}}{3}}$$

Practice 4

Given any five real numbers, show that at least two of them x and y satisfy the condition $|xy + 1| > |x - y|$.

Write each of these five numbers as $\tan \theta_k$ where $k = 1, 2, 3, 4, 5$, and $0 \leq \theta_k \leq 180°$

Dividing the range of $[0, 180°]$ into four sub-divisions: $[0, 45]$, $[45°, 90°]$, $[90°, 135°]$, and $[135°, 180°]$. By the pigeonhole principle[2], at least two of them must fall into one sub vision. Let them be θ_i and θ_j. Then

$$\left|\frac{x-y}{1+xy}\right| = \left|\frac{\tan \theta_i - \tan \theta_j}{1 + \theta_i \theta_j}\right| = |\tan(\theta_i - \theta_j)| \leq |\tan 45°| = 1$$

$$\therefore \quad |1 + xy| > |x - y|$$

Practice 5

Let $\{x_n\}$ and $\{y_n\}$ be two real number sequences which are defined as follow:

$$x_1 = y_1 = \sqrt{3}, \quad x_{n+1} = x_n + \sqrt{1 + x_n^2}, \quad y_{n+1} = \frac{y_n}{1 + \sqrt{1 + y_n^2}}$$

for all $n \geq 1$. Prove that $2 < x_n y_n < 3$ for all $n > 1$.

---

[2]The pigeonhole principle is discussed in the book *The Art of Thinking* written by the same author.

Chapter 8: Solutions

Let $x_n = \tan\alpha_n$ and $y_n = \tan\beta_n$ for all $n \geq 1$ where $\alpha_1 = \beta_1 = 60°$ and $0 < \alpha_n, \beta_n < 90°$ because all x_n and y_n are obviously positive. Then,

$$\begin{aligned}
\tan\alpha_{n+1} &= \tan\alpha_n + \sqrt{1 + \tan^2\alpha_n} \\
&= \tan\alpha_n + \sec\alpha_n \\
&= \frac{\sin\alpha_n + 1}{\cos\alpha_n} \\
&= \frac{1 - \cos(90° + \alpha_n)}{\sin(90° + \alpha_n)} \\
&= \tan\left(\frac{90° + \alpha_n}{2}\right) \\
&= \tan\left(90° - \frac{90° - \alpha_n}{2}\right)
\end{aligned}$$

The 2nd last step holds because of *(3.11)* on *page 23*.

$$\therefore \quad \alpha_n = 90° - \frac{90° - \alpha_{n-1}}{2} = \cdots = 90° - \frac{90° - \alpha_1}{2^{n-1}} = 90° - \frac{30°}{2^{n-1}} \quad (n > 1)$$

Let $\theta_n = \frac{30°}{2^{n-1}}$. Then,

$$x_n = \tan\alpha_n = \tan(90° - \theta_n) = \cot\theta_n \quad (n > 1)$$

Similarly, it can be shown that

$$y_n = \tan 2\theta_n \quad (n > 1)$$

This implies that for $n > 1$:

$$x_n y_n = \cot\theta_n \cdot \tan 2\theta_n = \frac{1}{\tan\theta_n} \cdot \frac{2\tan\theta_n}{1 - \tan^2\theta_n} = \frac{2}{1 - \tan^2\theta_n}$$

Because $0 < \theta_n < 30°$, therefore $0 < \tan^2\theta_n < \frac{1}{3}$. Substituting this range to the relation above immediately leads to the conclusion that

$$2 < x_n y_n < 3$$

Chapter 8: Solutions

Practice 6

Let x, y, z be three positive real numbers satisfying $xyz+x+z = y$. Find the maximum value of

$$P = \frac{2}{x^2+1} - \frac{2}{y^2+1} + \frac{3}{z^2+1}$$

First, we claim $xz \neq 1$. This is because if $xz = 1$, then we will have $x + y = 0$ which contradicts the condition that x and z are both positive. Hence, we can conclude

$$xyz + x + z = y \Leftrightarrow y = \frac{x+z}{1-xz}$$

Let $x = \tan\alpha$, $y = \tan\beta$ and $z = \tan\gamma$ where $\alpha, \beta, \gamma \in (0, 90°)$.

$$\tan\beta = \frac{\tan\alpha + \tan\gamma}{1 - \tan\alpha\tan\gamma} = \tan(\alpha+\gamma) \implies \beta = \alpha + \gamma$$

Then, we have

$$P = \frac{2}{\tan^2\alpha+1} - \frac{2}{\tan^2(\alpha+\gamma)+1} + \frac{3}{\tan^2\gamma+1}$$
$$= 2\cos^2\alpha - 2\cos^2(\alpha+\gamma) + 3\cos^2\gamma$$
$$= (1+\cos 2\alpha) - (1+\cos 2(\alpha+\gamma)) + 3\cos^2\gamma$$
$$= (\cos 2\alpha - \cos 2(\alpha+\gamma)) + 3\cos^2\gamma$$
$$= 2\sin\alpha\sin(2\alpha+\gamma) + 3(1-\sin^2\gamma)$$
$$\leq 2\sin\gamma + 3 - 3\sin^2\gamma$$
$$= -3\left(\sin\gamma - \frac{1}{3}\right)^2 + \frac{10}{3}$$
$$\leq \boxed{\frac{10}{3}}$$

Equality holds if and only if

$$\begin{cases}\sin(2\alpha+\gamma) = 1 \\ \sin\gamma = \frac{1}{3}\end{cases} \implies \begin{cases}\sin\gamma = \frac{1}{3} \\ 2\alpha+\gamma = \frac{\pi}{2}\end{cases} \implies \begin{cases}x = \frac{\sqrt{2}}{2} \\ y = \sqrt{2} \\ z = \frac{\sqrt{2}}{4}\end{cases}$$

Chapter 8: Solutions

Practice 7

Let m be a positive integer. Show that

$$\sin\frac{2}{\sqrt{m}} < \frac{2}{\sqrt{m+1}}$$

This can be proved using the same substitution as that used in *Example 7.3.2* on *page 80*.

When $m \leq 3$, the to-be-proved relation obviously holds because the right side is greater or equal to 1. When $m > 3$, let $m = \cot^2 \alpha$ where $\alpha \in (0, \frac{\pi}{2})$. Then

$$\sin\frac{2}{\sqrt{m}} < \frac{2}{\sqrt{m+1}}$$
$$\Leftrightarrow \quad \sin(2\tan\alpha) < 2\sin\alpha$$
$$\Leftrightarrow \quad \sin(\tan\alpha)\cos(\tan\alpha) < \sin\alpha$$

Because

$$\cot^2\alpha = m > 3 \implies \tan\alpha < \frac{\sqrt{3}}{3} < \frac{\pi}{2}$$

Therefore, we have $\sin\tan\alpha < \tan\alpha$. This means it is sufficient to show

$$\tan\alpha\cos(\tan\alpha) < \sin\alpha \quad \Leftrightarrow \quad \cos(\tan\alpha) < \cos\alpha$$

Because $0 < \alpha < \tan\alpha < \frac{\pi}{2}$, the last relation clearly holds.

Some of these steps utilize trigonometric functions' increasing and decreasing properties in the $(0, \pi/2)$ region as well as the inequality *(2.11)* on *page 12*.

Chapter 8: Solutions

Practice 8

Prove
$$\frac{1}{\sqrt{2019}} < \underbrace{\sin\sin\sin\cdots\sin}_{2017}\frac{\sqrt{2}}{2} < \frac{2}{\sqrt{2019}}$$

This problem can be directly derived from the conclusions of previous practice and *Example 7.3.2* on *page 80*.

$$\begin{aligned}
&\underbrace{\sin\sin\sin\cdots\sin}_{2017}\frac{\sqrt{2}}{2} \\
=\ &\underbrace{\sin\sin\sin\cdots\sin}_{2017}\frac{1}{\sqrt{2}} \\
>\ &\underbrace{\sin\sin\sin\cdots\sin}_{2016}\frac{1}{\sqrt{3}} \\
>\ &\underbrace{\sin\sin\sin\cdots\sin}_{2015}\frac{1}{\sqrt{4}} \\
>\ &\cdots \\
>\ &\sin\frac{1}{\sqrt{2018}} \\
>\ &\frac{1}{\sqrt{2019}}
\end{aligned}$$

$$\begin{aligned}
&\underbrace{\sin\sin\sin\cdots\sin}_{2017}\frac{\sqrt{2}}{2} \\
=\ &\underbrace{\sin\sin\sin\cdots\sin}_{2017}\frac{2}{\sqrt{8}} \\
<\ &\underbrace{\sin\sin\sin\cdots\sin}_{2016}\frac{2}{\sqrt{9}} \\
<\ &\underbrace{\sin\sin\sin\cdots\sin}_{2015}\frac{2}{\sqrt{10}} \\
<\ &\cdots
\end{aligned}$$

$$< \sin\frac{2}{\sqrt{2024}}$$
$$< \frac{2}{\sqrt{2025}}$$
$$< \frac{2}{\sqrt{2019}}$$

Practice 9

Solve this equation

$$2\sqrt{2}x^2 + x - \sqrt{1-x^2} - \sqrt{2} = 0$$

Clearly, it must be true that $|x| \leq 1$ in order to make the term $\sqrt{1-x^2}$ defined. Therefore, we can let $x = \sin\theta$ where $\theta \in [-\frac{\pi}{2}, \frac{\pi}{2}]$. Setting this to the given equation yields

$$2\sqrt{2}\sin^2\theta + \sin\theta - \cos\theta - \sqrt{2} = 0$$
$$\sqrt{2}(2\sin^2\theta - 1) + (\sin\theta - \cos\theta) = 0$$
$$\sqrt{2}(\sin^2\theta - \cos^2\theta) + (\sin\theta - \cos\theta) = 0$$
$$\sqrt{2}(\sin\theta + \cos\theta)(\sin\theta - \cos\theta) + (\sin\theta - \cos\theta) = 0$$
$$(\sin\theta - \cos\theta)(\sqrt{2}(\sin\theta + \cos\theta) + 1) = 0$$
$$(\sqrt{2}\sin(\theta - 45°))(2\sin(\theta + 45°) + 1) = 0$$

Hence, we find two solutions:

$$\sin(\theta - 45°) = 0 \implies \theta = 45° \implies x = \boxed{\frac{\sqrt{2}}{2}}$$

$$\sin(\theta + 45°) = -\frac{1}{2} \implies \theta = -75° \implies x = \boxed{-\frac{\sqrt{6} + \sqrt{2}}{4}}$$

Chapter 8: Solutions

Practice 10

Let a and b be two positive real numbers not exceeding 1. Prove

$$\frac{1}{\sqrt{a^2+1}} + \frac{1}{\sqrt{b^2+1}} \leq \frac{2}{\sqrt{1+ab}}$$

(Russia)

Let $a = \tan\alpha$ and $b = \tan\beta$ where $\alpha, \beta \in (0, \frac{\pi}{4}]$. Then

$$\begin{aligned}
& \frac{1}{\sqrt{a^2+1}} + \frac{1}{\sqrt{b^2+1}} \leq \frac{2}{\sqrt{1+ab}} \\
\Leftrightarrow\quad & \frac{1}{\sqrt{\tan^2\alpha+1}} + \frac{1}{\sqrt{\tan^2\beta+1}} \leq \frac{2}{\sqrt{1+\tan\alpha\tan\beta}} \\
\Leftrightarrow\quad & \cos\alpha + \cos\beta \leq 2\sqrt{\frac{\cos\alpha\cos\beta}{\cos\alpha\cos\beta+\sin\alpha\sin\beta}} \\
\Leftrightarrow\quad & \cos\alpha + \cos\beta \leq 2\sqrt{\frac{\cos\alpha\cos\beta}{\cos(\alpha-\beta)}}
\end{aligned}$$

Because $\alpha, \beta \in (0, \frac{\pi}{4}]$, it must hold that

$$0 < \cos\alpha, \cos\beta, \cos(\alpha-\beta) < 1$$

Hence, the last inequality is equivalent to its square, i.e.:

$$\begin{aligned}
& (\cos\alpha + \cos\beta)^2 \leq 4 \cdot \frac{\cos\alpha\cos\beta}{\cos(\alpha-\beta)} \\
\Leftrightarrow\quad & \cos^2\alpha + \cos^2\beta + 2\cos\alpha\cos\beta \leq \frac{4\cos\alpha\cos\beta}{\cos(\alpha-\beta)} \\
\Leftrightarrow\quad & \cos(\alpha-\beta)(\cos^2\alpha + \cos^2\beta) \leq (4 - 2\cos(\alpha-\beta))\cos\alpha\cos\beta
\end{aligned}$$

Because $0 < \cos(\alpha-\beta) < 1$, therefore it is sufficient to show

$$\begin{aligned}
& \cos(\alpha-\beta)(\cos^2\alpha + \cos^2\beta) \leq (4-2)\cos\alpha\cos\beta \\
\Leftrightarrow\quad & \cos(\alpha-\beta)(\cos^2\alpha + \cos^2\beta) \leq 2\cos\alpha\cos\beta \\
\Leftrightarrow\quad & \cos(\alpha-\beta)\left(\frac{1+\cos 2\alpha}{2} + \frac{1+\cos 2\beta}{2}\right) \leq 2\cos\alpha\cos\beta \\
\Leftrightarrow\quad & \cos(\alpha-\beta)(\cos 2\alpha + \cos 2\beta + 2) \geq 4\cos\alpha\cos\beta
\end{aligned}$$

Applying sum-to-product transformation on the left gives

$$\cos(\alpha-\beta)(2\cos(\alpha+\beta)\cos(\alpha-\beta) + 2)$$

Applying product-to-sum transformation on the right gives

$$2(\cos(\alpha+\beta)+\cos(\alpha-\beta))$$

Setting these back and rearranging yields

$$\Leftrightarrow \quad \cos^2(\alpha-\beta)\cos(\alpha+\beta) \ \leq \ \cos(\alpha+\beta)$$

This clearly holds because $\cos^2(\alpha-\beta) \leq 1$ and

$$0 < \alpha, \beta \leq \frac{\pi}{4} \implies 0 < \alpha+\beta \leq \frac{\pi}{2} \implies 0 \leq \cos(\alpha+\beta) < 1$$

Chapter 8: Solutions

www.ingramcontent.com/pod-product-compliance
Lightning Source LLC
Chambersburg PA
CBHW070246230526
45470CB00002B/494